STUDY GUIDE

for the Fourth Edition of

Aircraft Rescue and Fire Fighting

WRITTEN BY
John Joerschke

EDITED BY
Barbara Adams

Published by
Fire Protection Publications
Oklahoma State University
Stillwater, Oklahoma

The International Fire Service Training Association

The International Fire Service Training Association (IFSTA) was established in 1934 as a "nonprofit educational association of fire fighting personnel who are dedicated to upgrading fire fighting techniques and safety through training." To carry out the mission of IFSTA, Fire Protection Publications was established as an entity of Oklahoma State University. Fire Protection Publications' primary function is to publish and disseminate training texts as proposed and validated by IFSTA. As a secondary function, Fire Protection Publications researches, acquires, produces, and markets high-quality learning and teaching aids consistent with IFSTA's mission.

NOTICE: The questions in this study guide are taken from the information presented in the fourth edition of *Aircraft Rescue and Fire Fighting,* an IFSTA-validated manual. The questions are *not validated test questions and are not intended to be duplicated or used for certification or promotional examinations;* this guide is intended to be used as a tool for studying the information presented in *Aircraft Rescue and Fire Fighting.*

Copyright © 2001 by the Board of Regents, Oklahoma State University

All Rights reserved. No part of this publication may be reproduced without prior written permission from the publisher.

ISBN 0-87939-200-2

Fourth Edition
First Printing, October 2001

Printed in the United States of America

If you need additional information concerning our organization or assistance with manual orders, contact:
Customer Service, Fire Protection Publications, Oklahoma State University
930 N. Willis, Stillwater, OK 74078-8045
1-800-654-4055   FAX: 405-744-8204

For assistance with training materials, to recommend material for inclusion in an IFSTA manual, or to ask questions on manual content, contact:
Editorial Department, Fire Protection Publications, Oklahoma State University
930 N. Willis, Stillwater, OK 74078-8045
405-744-5723   FAX: 405-707-0024   E-mail: editors@ifstafpp.okstate.edu

Oklahoma State University in compliance with Title VI of the Civil Rights Act of 1964 and Title IX of the Educational Amendments of 1972 (Higher Education Act) does not discriminate on the basis of race, color, national origin or sex in any of its policies, practices or procedures. This provision includes but is not limited to admissions, employment, financial aid and educational services.

How to Use this Book

This study guide is developed to be used in conjunction with and as a supplement to the first edition of the IFSTA manual, **Aircraft Rescue and Fire Fighting.** The questions in this guide are designed to help you remember information and to make you think—they are *not* intended to trick or mislead you. To derive the maximum learning experience from these materials, use the following procedure:

Step 1: Read one chapter at a time in the **Aircraft Rescue and Fire Fighting** manual. After reading the chapter, underline or highlight important terms, topics, and subject matter in that chapter.

Step 2: Open the study guide to the corresponding chapter. Answer all of the questions in the study guide for that chapter. You may have to refer to a dictionary or the glossary in the **Fire Service Orientation and Terminology (O&T)** manual for terms that appear in context but are not defined. After you have defined terms and answered all questions possible, check your answers with those in the answer section at the end of the study guide.

✓**Note:** *Do not* answer each question and then immediately check the answer for the correct response.

If you find that you have answered any question incorrectly, find the explanation of the answer in the **Aircraft Rescue and Fire Fighting** manual. The number in parentheses after each answer in the answer section identifies the page on which the answer or term can be found. Correct any incorrect answers, and review material that was answered incorrectly.

Step 3: Go to the next chapter of the manual and repeat Steps 1 and 2.

Chapter 1: Qualifications for Aircraft Rescue and Fire Fighting Personnel

True/False

A. Write *True* or *False* before each of the following statements. Correct those statements that are false.

_____ 1. Federal Aviation Administration Advisory Circulars outline legal requirements.

_____ 2. ARFF personnel may respond to medical emergencies and therefore should have appropriate emergency medical care training.

_____ 3. Before being certified as an ARFF firefighter, the candidate must also be certified as a Fire Fighter I and II.

_____ 4. Classroom training and looking at maps provide all the information a firefighter needs to become familiar with an airport.

_____ 5. All airports have target hazard areas that the airport firefighter must be able to identify.

_____ 6. Firefighters are not responsible for airport security.

1

_____ 7. Airport firefighters must be totally familiar with all aspects of operating the ARFF vehicles at their airports.

_____ 8. Because ARFF vehicles are designed to deliver mass quantities of water/agent, agent management is not a concern in ARFF operations.

_____ 9. Most apparatus or equipment failures are unavoidable.

_____ 10. All aircraft fires require the use of a Class B extinguishing agent.

_____ 11. A well-trained team of firefighters with an attack plan and an adequate amount of extinguishing agent, properly applied, can contain most fires in their early stages.

_____ 12. An airport emergency plan should be as complete and detailed as necessary to ensure that all involved agencies are aware of their roles and responsibilities under various conditions.

_____ 13. Federal regulations prohibit air transportation of inherently hazardous cargoes.

_____ 14. All ARFF personnel must be trained in accordance with plans and procedures for safely handling haz mat incidents.

_____ 15. Emergency response organizations have a moral obligation to provide a safe working atmosphere for their employees.

Multiple Choice

B. Write the letter of the best answer on the blank before each statement.

_____ 1. The *Standard for Airport Fire Fighter Professional Qualifications* is _____.
   a. NFPA 1002
   b. NFPA 1003
   c. NFPA 1030
   d. NFPA 2001

_____ 2. Prospective candidates for ARFF training programs must meet the medical requirements of _____.
   a. NFPA 1582
   b. NFPA 1825
   c. NFPA 2008
   d. NFPA 2111

_____ 3. The recommended performance criteria for maintaining proficiency and effective aircraft rescue and fire fighting services are contained in _____.
   a. NFPA 104
   b. NFPA 405
   c. NFPA 1050
   d. NFPA 4001

_____ 4. Which of the following comprises proficiency training conducted within the firefighter's own fire department?
   a. Cadet (recruit) training
   b. On-the-job training
   c. In-service training
   d. Special courses and seminars

_____ 5. Extension programs, conferences, short courses, and workshops are examples of _____.
   a. Cadet (recruit) training
   b. On-the-job training
   c. In-service training
   d. Special courses and seminars

_____ 6. NFPA 472 outlines the requirements for responders to _____.
   a. Hazardous materials incidents
   b. Aircraft fires
   c. Medical emergencies
   d. Airplane hijackings

_____ 7. NFPA 1003 includes requirements for all of the following major duties *except* _____.
   a. Response
   b. Fire suppression
   c. Rescue
   d. Vehicle requirements

Aircraft Rescue and Fire Fighting

1

_____ 8. Breakaway fences and gates are also known as _____.
   a. Easy-give
   b. Fractal
   c. Frangible
   d. Drive-through

_____ 9. Airport firefighters must be familiar with aircraft construction features and materials as they relate to _____.
   a. Aerodynamics
   b. Forcible entry
   c. Maneuverability
   d. Airworthiness

_____ 10. Which of the following statements about firefighter safety is true?
   a. From an ARFF standpoint, aircraft incidents will always involve a fire.
   b. Small incidents and daily operations should be handled without initiating an incident management system.
   c. Knowing how to don and use protective equipment is more important than knowing its limitations.
   d. The department should have a critical incident stress debriefing program in place.

_____ 11. ARFF communications include all of the following *except* methods by which _____.
   a. Information is provided to the public
   b. The telecommunications center is notified of an emergency
   c. The communications center can notify the proper fire fighting forces
   d. Information is exchanged at the scene

_____ 12. Which of the following statements about ARFF communications is true?
   a. The firefighter must know how to use the communication center's alarm-receiving equipment.
   b. Mutual aid companies must use the same radio frequencies as the airport fire department.
   c. ARFF vehicles responding to emergencies are exempt from requirements to receive clearance before entering any area of the airport.
   d. All communications between ARFF personnel and the pilot of an emergency aircraft must be channeled through the air traffic control tower.

_____ 13. All of the following statements about aircraft rescue tools and equipment are true *except* _____.
   a. Selection and use of the appropriate tools are imperative to gaining access into an aircraft.
   b. Which tools are actually needed depends on the skills of the firefighter.
   c. A thorough understanding of the available tools ensures that the firefighter can perform forcible entry tasks efficiently and safely.
   d. Forcible entry is a natural talent that begins with strength and agility.

____ 14. Who is responsible for ensuring that an apparatus and the equipment it carries are ready at all times?
   a. Crew
   b. Chief mechanic
   c. Driver/operator
   d. Airport director

____ 15. The *Standard for Fire Apparatus Driver/Operator Professional Qualifications* is ____.
   a. NFPA 1002
   b. NFPA 1003
   c. NFPA 1004
   d. NFPA 1005

____ 16. Which of the following statements about aircraft rescue operations is true?
   a. Firefighters should apply extinguishing agents to the passenger compartment *only* if it is involved in fire.
   b. Extinguishment is a higher priority than evacuation.
   c. Evacuation may require nothing more than opening the aircraft's normal entry doors.
   d. Exit openings must be protected by turrets flowing agent to provide a foam blanket.

____ 17. To obtain information on hazardous materials/dangerous goods for a given situation, firefighters must be able to use the ____.
   a. NFPA *Standard for the Identification of Hazardous Materials*
   b. DOT *Emergency Response Guidebook*
   c. UL *Dangerous Substances Tables*
   d. ISO *Hazardous Materials Handbook*

____ 18. The principal providers of immediate technical information about hazardous materials in the United States and Canada respectively are ____.
   a. DOT and TC
   b. CHEMTREC and CANUTEC
   c. OSHA and CHSO
   d. SAE-Tech and ISO

Identify

C. Identify the following abbreviations associated with qualifications for aircraft rescue and fire fighting personnel. Write the correct interpretation before each.

_____ 1. ARFF

_____ 2. NFPA

_____ 3. CFR

_____ 4. FAR

_____ 5. ICAO

_____ 6. FAA

1

_____ 7. AC

_____ 8. APU

_____ 9. DFDR

_____ 10. CVR

_____ 11. PPE

_____ 12. IMS

_____ 13. CISD

_____ 14. ERG

_____ 15. CHEMTREC

_____ 16. CANUTEC

Chapter 2 Airport Familiarization

Matching

A. Match to their definitions terms associated with airports. Write the appropriate letters on the blanks.

____ 1. Flight path at right angles to the landing runway off its upwind leg

____ 2. Flight path parallel to the landing runway in the direction opposite to landing

____ 3. Flight path at a right angle to the landing runway off the approach end

____ 4. Portion of the landing pattern in which the aircraft is lined up with the runway and is heading straight in to land

____ 5. Specially designated and prepared surface on an airport for aircraft to travel to and from runways, hangars, etc.

____ 6. Airport marking that can be crossed only with permission of the control tower

____ 7. The features of the earth's surface, both natural and constructed, and the relationships among them

____ 8. Area where entry is limited in order to eliminate unnecessary or unauthorized traffic

____ 9. Predetermined location for temporary parking of aircraft experiencing problems with hazardous cargo

a. Base leg
b. Topography
c. Crosswind leg
d. Controlled access point
e. Final approach
f. Downwind leg
g. Designated isolation area
h. ILS critical area
i. Hold bar
j. Taxiway

2

True/False

B. Write *True* or *False* before each of the following statements. Correct those statements that are false.

1. _____ Emergency response personnel must be able to find their way quickly to any point on the airport—even at night or when visibility is reduced by inclement weather.

2. _____ Firefighters assigned to nearby structural stations must be familiar with airport layout, driving regulations, and communications procedures.

3. _____ ARFF vehicles are prohibited from using the same access routes as aircraft.

4. _____ Taxiway designations are standardized at all airports on the North American continent.

5. _____ No vehicle or aircraft is allowed close to the ILS area when it is in operation for an aircraft landing.

6. _____ Aircraft always have the right of way over ARFF apparatus.

7. _____ The CRFFAA's length extends 500 feet *(152 m)* from each end of the runway.

_____ 8. The topography in the immediate area of a fire affects both its intensity and its rate and direction of spread.

_____ 9. Airport firefighters need to know the details of how navigational aids work.

_____ 10. Radio waves produced by some navigational aids may be harmful to firefighters.

_____ 11. ARFF crews should know the load limits of all bridges on the airport and in the local vicinity.

_____ 12. Airport firefighters should rely on meteorological instruments—**not** try to visually monitor weather conditions.

_____ 13. If necessary, ARFF vehicles may be used to strike sections of a frangible fence or gate.

_____ 14. Distribution of water from fixed water supply systems must *not* be through domestic water supply mains.

2

_____ 15. All firefighters assigned to an airport should periodically visit each fuel storage and distribution site and become thoroughly familiar with its location, function, and operation.

_____ 16. NFPA standards require grounding to a static ground electrode in the pavement during fueling operations.

_____ 17. Open flames should be strictly controlled or prohibited within 25 feet *(8 m)* of any aircraft fueling operation.

_____ 18. Defueling operations can be just as hazardous as fueling operations.

_____ 19. Multipurpose dry chemical fire extinguishers are recommended for use during aircraft fueling operations.

_____ 20. All airports have the same type of standardized system to control spilled fuels.

Multiple Choice

C. **Write the letter of the best answer on the blank before each statement.**

___ 1. Aircraft are normally directed to take off and land _____.
   a. With the wind
   b. Into the wind
   c. Perpendicular to the wind
   d. Without regard to wind direction

10
Aircraft Rescue and Fire Fighting

2. An airport that has an operating tower with air traffic controllers is called a _____ airport.
    a. Secured
    b. Directed
    c. Supervised
    d. Controlled

3. The FAA classifies airports used by air carrier aircraft with seating for more than 30 passengers according to the _____.
    a. Average number of passengers departing daily and average number of total daily departures of all aircraft
    b. Length of the air carrier aircraft and average number of total daily departures of all aircraft
    c. Length of the air carrier aircraft and average number of passengers departing daily
    d. Length of the air carrier aircraft and average number of daily departures of air carrier aircraft

4. The minimum designated FAA index for an airport used by air carrier aircraft is _____.
    a. Index A
    b. Index D
    c. Index E
    d. Index F

5. The ICAO bases its airport categories on the _____.
    a. Total daily passenger capacity of air carrier aircraft using the airport
    b. Longest aircraft using the airport and their fuselage width
    c. Longest aircraft using the airport and their average passenger capacity
    d. Longest aircraft using the airport without regard to other factors

6. How many fire fighting vehicles does the FAA require at a Class C airport?
    a. One
    b. One or two
    c. Two or three
    d. Three

7. A runway designated 27 for aircraft approaching from the east would be designated _____ for aircraft approaching from the west.
    a. 09
    b. 153
    c. 207
    d. 18

8. At most airports a runway designated 34L would be parallel to a runway designated _____.
    a. 16L
    b. 33L
    c. 34K
    d. 34R

9. At most airports a runway designated 18C would run between runways designated _____.
    a. 18B and 18D
    b. 90C and 270C
    c. 17C and 19C
    d. 18L and 18R

_____ 10. What color lights outline taxiways?
   a. Green
   b. White
   c. Amber
   d. Blue

_____ 11. What color lights designate runways?
   a. Green
   b. White
   c. Amber
   d. Red

_____ 12. What color lights designate the approach end of runways?
   a. Green
   b. White
   c. Amber
   d. Blue

_____ 13. What color lights mark obstructions or the departure end of runways?
   a. Green
   b. Red
   c. Amber
   d. Blue

_____ 14. What color lights designate hold bars?
   a. Green
   b. White
   c. Amber
   d. Blue

_____ 15. What color lights designate the last 2,000 feet *(610 m)* at the departure end of runways?
   a. Green
   b. Red
   c. Amber
   d. Blue

_____ 16. What color marking is used for runway identification numbers/letters, landing zone bars, and striping?
   a. Yellow
   b. Green
   c. White
   d. Red

_____ 17. What color markings designate taxiways and Instrument Landing System critical areas?
   a. Yellow
   b. Green
   c. White
   d. Red

_____ 18. Yellow chevrons pointing at the displaced threshold line indicate that the area is _____.
   a. To be used for taxi and rollout
   b. An ILS critical area
   c. Closed to aircraft and vehicle operations
   d. Not suitable for aircraft operations

_____ 19. Which type of airport sign has white inscriptions on a red background?
   a. Location sign
   b. Runway hold position sign
   c. Direction sign
   d. Information sign

2

_____ 20. The sign in the figure below indicates _____.

a. ILS critical area boundary
b. No entry
c. RSA/OFZ boundary
d. Taxiway ending

_____ 21. The sign in the figure below indicates _____.

a. ILS critical area boundary
b. No entry
c. RSA/OFZ boundary
d. Taxiway ending

_____ 22. Which type of airport sign has black inscriptions on a yellow background?
a. Location sign
b. Runway hold position sign
c. Direction sign
d. Mandatory instruction sign

_____ 23. A white number *3* on a black background indicates _____.
a. 300 meters of remaining runway
b. 3,000 meters of remaining runway
c. 300 feet of remaining runway
d. 3,000 feet of remaining runway

_____ 24. The key factor in determining the most appropriate response route for ARFF vehicles is _____.
a. Apparatus type
b. Incident type
c. Airport layout
d. Wind direction

_____ 25. A type of map marked with rectangular coordinates or azimuthal bearings using polar coordinates is called a _____.
a. Topographic map
b. Grid map
c. Graphing map
d. Orthographic map

_____ 26. In an emergency situation, the location of the emergency scene should be described primarily in terms of _____.
a. The distance and direction from the control tower
b. Landmarks
c. The distance and direction from the fire department
d. The grid system

Aircraft Rescue and Fire Fighting

2

___ 27. Major concerns for ARFF personnel responding to an emergency at an airport terminal typically include all of the following *except* ___.
   a. Hazardous materials storage areas
   b. Life safety
   c. Jetways
   d. Baggage handling areas

___ 28. Major concerns for ARFF personnel responding to an emergency at an airport maintenance facility typically include all of the following *except* ___.
   a. Offices and record keeping areas
   b. Jetways
   c. Use of flammable and hazardous chemicals
   d. Welding, cutting, and grinding operations

___ 29. Who usually dictates the size and number of fire extinguishers required in an aircraft maintenance facility?
   a. FAA
   b. NFPA
   c. Airport director
   d. Local AHJ

___ 30. The NFPA *Standard on Aircraft Hangars* is ___.
   a. NFPA 10
   b. NFPA 109
   c. NFPA 409
   d. NFPA 490

___ 31. Confined spaces and high-voltage switching equipment are typically hazards in ___.
   a. Utility structures and vaults
   b. Control towers
   c. Multilevel parking structures
   d. Hotels/motels, stores, and restaurants

___ 32. Low overhead clearances and weight restrictions are typically hazards in ___.
   a. Utility structures and vaults
   b. Control towers
   c. Multilevel parking structures
   d. Hotels/motels, stores, and restaurants

___ 33. The most congested areas of airports tend to be ___.
   a. Parking lots
   b. Ramps and aprons
   c. Terminal hubs
   d. Access roads

___ 34. The maneuver in which a passenger aircraft backs away from the jetway or terminal area is called a ___.
   a. Pushback
   b. Backout
   c. Pullout
   d. Rollback

____ 35. To monitor ground activities through direct observation, airport fire stations should be ____.
   a. In an isolated location with an unobstructed view
   b. Located at either end of the airport
   c. Equipped with surveillance cameras
   d. In a central location

____ 36. The number-one fire-prevention consideration at airports is ____.
   a. Aircraft takeoff and landing
   b. Careless smoking in terminals
   c. Aircraft maintenance and repair
   d. Fueling operations

____ 37. The most common method of aircraft fuel delivery is by a ____.
   a. Subsurface fuel hydrant   c. Fueling island
   b. Tank truck   d. Aboveground pipe system

____ 38. The automatic release mechanism that shuts down the fueling operation in an emergency is called a ____.
   a. Safety valve   c. Quick stop
   b. Dead man   d. Cripple

____ 39. Smaller aircraft predominantly use the ____ method of fueling.
   a. Single-point   c. Over-the-wing
   b. Onboard   d. Underbelly

____ 40. Fuel vapors are ____.
   a. Invisible but lighter than air   c. Visible but lighter than air
   b. Invisible but heavier than air   d. Visible but heavier than air

____ 41. Which of the following statements regarding static electricity is true?
   a. The generation of static charges increases when humidity is high.
   b. Bonding should be done only on equipment connected to painted metallic surfaces.
   c. Silk is less prone than cotton to produce static electricity.
   d. The degree of static buildup depends on the velocity of the fuel's movement.

____ 42. Mobile radio equipment and cellular telephones produce ____, which is/are hazardous in close proximity to fueling operations.
   a. Electromagnetic energy   c. Electromechanical sparks
   b. Static electricity   d. Radiation

____ 43. Fire extinguishers with a minimum rating of ____ should be readily accessible within the area of fueling operations.
   a. 20-B   c. 40-B
   b. 30-B   d. 50-B

2

___ 44. For normal ramp area protection, the maximum distance between fire extinguishers should be ___.
   a. 75 feet *(23 m)*
   b. 100 feet *(30 m)*
   c. 125 feet *(38 m)*
   d. 150 feet *(46 m)*

___ 45. The final interceptor/separator for an airport's entire drainage system should be designed to dispose of combustible or flammable liquids into ___.
   a. The sewage system
   b. Fuel recycling tanks
   c. Containment facilities
   d. Fuel/water separators

Identify

D. **Identify the following abbreviations associated with airports. Write the correct interpretation before each.**

_____ 1. AHJ

_____ 2. ILS

_____ 3. ATC

_____ 4. CRFFAA

_____ 5. NAVAID

_____ 6. GPU

_____ 7. FOD

_____ 8. SIDA

Chapter 3 Aircraft Familiarization

Matching

A. Match to their definitions terms associated with aircraft. Write the appropriate letters on the blanks.

____ 1. Main body of an aircraft to which the wings and tail are attached

____ 2. Develops the major portion of the lift required for flight

____ 3. Includes the horizontal and vertical stabilizers, rudders, and elevators

____ 4. Housing around an externally mounted aircraft engine

____ 5. Controls the rolling (banking) motion of the aircraft

____ 6. Controls the up-and-down pitch motion of the aircraft

____ 7. Controls the yaw, or turning motion, of the aircraft

____ 8. Airfoil that extends from the leading edge and/or trailing edge of a wing

____ 9. Airfoil that extends only from the leading edge of a wing

____ 10. Movable panel on the upper surface of a wing

____ 11. Provides lift and propulsion for a helicopter

____ 12. Provides directional control for a helicopter

a. Flap
b. Tail rotor
c. Nacelle
d. Elevator
e. Fuselage
f. Speed brake
g. Spoiler
h. Tail
i. Slat
j. Main rotor
k. Aileron
l. Wing
m. Rudder

3

True/False

B. Write *True* or *False* before each of the following statements. Correct those statements that are false.

_____ 1. Depending on their uses, some aircraft may be included in more than one category.

_____ 2. Many freighters are former passenger aircraft modified to carry cargo.

_____ 3. In the military aircraft designation *A-10*, the letter *A* indicates that the aircraft belongs to the Air Force.

_____ 4. Fire-fighting aircraft may be used for medevac and high-angle rescue roles in addition to transporting smokejumpers and dropping fire-fighting agent.

_____ 5. The internal fuel tanks in rotary-wing aircraft are usually located behind the cargo compartment.

_____ 6. Conventional landing gear on fixed-wing aircraft consists of a single strut under the nose and two main struts extending from under the wings or out of the fuselage.

_____ 7. Helicopters with skids often "hover taxi" to move along taxiways or the parking ramp.

_____ 8. Reciprocating engines in aircraft are air-cooled to eliminate weight.

_____ 9. Disconnecting the battery prevents an internal-combustion reciprocating aircraft engine from starting.

_____ 10. Jet blast can easily upset any vehicle that is driven too close to the rear of an operating jet engine.

_____ 11. When a number of jet engines are operating in an area, ground personnel can easily tell which engines are operating and which are not.

_____ 12. Ballistic recovery systems use an ejection device to quickly deploy a parachute during catastrophic emergencies.

_____ 13. The FAA considers stowage compartments such as overhead storage areas for carry-on articles and baggage as cargo compartments.

_____ 14. All new aircraft must be equipped with Class D cargo compartments.

_____ 15. In lieu of providing extinguishment, a Class E cargo compartment must have a means of shutting off the flow of ventilating air to or within the compartment.

3

_____ 16. The best device for manually opening a mechanically operated cargo door is a pneumatic driver.

_____ 17. Aircraft fire-fighting operations may be affected both by the inherent properties of the aircraft construction materials and by the manner in which these components are assembled.

_____ 18. One of aluminum's advantages for aircraft construction is that it withstands high heat well.

_____ 19. From a fire-fighting standpoint, the most comprehensive study of composite materials in aircraft is "Advanced Composites/Advanced Aerospace Materials (AC/AAM): Mishap Risk Control and Mishap Response" by John M. Olson.

_____ 20. The largest system in the aircraft is the fuel system.

_____ 21. The significance of damage to the aircraft's exterior is a reliable indicator of the likelihood of damage to the fuel system.

_____ 22. Aircraft manufacturers tend to adopt improvements in fuel-tank technology as soon as they become available.

_____ 23. Normal repair procedures call for maintenance personnel to enter the confined spaces of fuel tanks.

_____ 24. Aircraft with engines in the tail section may have fuel lines routed through interior walls, through the roof, or between the main cabin floor and the cargo area.

_____ 25. Most all aircraft rims in a wheel assembly have fusible plugs that melt when the rim reaches a certain temperature, automatically deflating the tire.

_____ 26. Brakes and wheels reach their maximum temperatures as the aircraft comes to a stop.

_____ 27. Aircraft batteries have a capacity only one-third that of the average automobile battery.

_____ 28. The electrical system should be de-energized before proceeding with aircraft shutdown functions.

_____ 29. GPUs can be used to produce either AC or DC power and come in battery-powered or gas-fueled models.

3

_____ 30. Once the reaction is started in a chemically generated oxygen system, it is impossible to stop until the unit has exhausted its chemical.

_____ 31. When LOX has been spilled on asphalt, it is so explosively unstable that even walking on it may cause a violent reaction.

_____ 32. One acceptable way to stop a liquid oxygen leak is to spray the leak with aqueous film forming foam.

_____ 33. Once the battery has been disconnected and all electrical power removed, an aircraft's fire suppression system will not operate.

_____ 34. The FAA requires all pressurized cylinders on aircraft to have pressure-relief valves.

_____ 35. Shutdown procedures in military aircraft are often the same as in similar commercial aircraft.

_____ 36. Escape slides may be disconnected from the aircraft and used as rafts.

_____ 37. The aircraft's so-called "black boxes" are actually painted yellow.

_____ 38. If a military aircraft functions in a dual role, it carries the letter for its more frequent assignment.

_____ 39. On a military aircraft with a hydrazine-powered EPU, personnel must safety the EPU before de-energizing the electrical power.

_____ 40. The exit doors and hatches on all military aircraft operate on the same basic system.

_____ 41. Ejection systems are *not* normally found in training aircraft.

_____ 42. Hands-on training is the only way to become competent and confident in emergency procedures involving ejection systems.

_____ 43. Aircraft canopies generally weigh less than 50 pounds *(23 kg)*.

_____ 44. The catapult for an ejection seat may hurl a 300-pound *(136 kg)* object at an initial rate of 60 feet *(20 m)* per second.

_____ 45. If a cast high explosive melts and runs from a ruptured ammunition case, it becomes extremely sensitive to shock until it resolidifies, at which point it becomes inert.

3

_____ 46. Handling incidents involving nuclear weapons is the responsibility of military fire suppression personnel who have received specific training and guidance on these weapons systems.

Multiple Choice

C. Write the letter of the best answer on the blank before each statement.

_____ 1. Generally, aircraft are categorized according to their _____.
   a. Size
   b. Engine type
   c. Intended purpose
   d. Manufacturer

_____ 2. The maximum number of passengers that narrow-body aircraft can seat is _____.
   a. 155
   b. 185
   c. 235
   d. 255

_____ 3. Federal aviation regulations require any aircraft with a doorsill height of 6 feet *(2 m)* or more off the ground to be equipped with _____.
   a. An emergency escape slide
   b. Clearance-height signs
   c. A retractable landing gear
   d. Retractable stairs

_____ 4. Wide-body aircraft may carry over _____ of jet fuel.
   a. 13,000 gallons *(52 000 L)*
   b. 25,000 gallons *(95 000 L)*
   c. 44,000 gallons *(165 000 L)*
   d. 58,000 gallons *(220 000 L)*

_____ 5. Wide-body aircraft may carry over _____ passengers.
   a. 400
   b. 450
   c. 500
   d. 550

_____ 6. New large aircraft may incorporate a passenger capacity of up to _____ passengers.
   a. 600
   b. 700
   c. 800
   d. 900

_____ 7. Which of the following statements about commuter/regional aircraft is true?
   a. The trend is away from the use of jet-powered aircraft in this role.
   b. There is usually no flight attendant on aircraft with fewer than 30 passengers.
   c. Few of these aircraft are pressurized.
   d. These aircraft must have at least two doors.

____ 8. Aircraft that carry passengers and cargo on the main deck and additional cargo below the deck are called _____.
   a. Compartmentalized aircraft
   b. Combi-aircraft
   c. Over-and-under aircraft
   d. Dual-purpose aircraft

____ 9. On narrow-body cargo freighters, the lower compartments are usually bulk-loaded with packages no heavier than _____ each.
   a. 40 pounds *(18 kg)*
   b. 55 pounds *(25 kg)*
   c. 70 pounds *(32 kg)*
   d. 85 pounds *(39 kg)*

____ 10. Which type of aircraft accounts for most aircraft-related fatalities?
   a. Commercial transport
   b. Commuter
   c. General aviation
   d. Business/corporate aviation

____ 11. Along with general aviation aircraft, which type of aircraft accounts for the largest variety of aircraft styles and configurations?
   a. Commercial transport
   b. Commuter
   c. Cargo
   d. Business/corporate aviation

____ 12. Which of the following statements about rotary-wing fire-fighting aircraft is true?
   a. They can carry a maximum of 500 gallons *(1 893 L)* of agent.
   b. They can be used to transport firefighters but not cargo.
   c. If they are used in wildland backfiring operations, they may carry flammable "ping pong balls" in the cargo area.
   d. Their use as infrared imaging platforms has proven ineffective.

____ 13. If involved in an accident, helicopters tend to _____.
   a. Collapse
   b. Disintegrate
   c. Remain rigid
   d. Blow apart

____ 14. Another name for an aircraft's tail is _____.
   a. Fin
   b. Empennage
   c. Nacelle
   d. Appendage

____ 15. On airliners, the cockpit is also referred to as the _____.
   a. Flight deck
   b. Cabin
   c. Captain's quarters
   d. Pilot's roost

____ 16. The general term for devices that enable the pilot to control the direction of flight is _____.
   a. Directional plane
   b. Aileron
   c. Rudder
   d. Flight control surface

3

____ 17. Most aircraft with internal-combustion reciprocating engines are used primarily for ____.
   a. Cargo aircraft
   b. Business/corporate aviation
   c. General aviation
   d. Commuter/regional aircraft

____ 18. Aircraft powered by internal-combustion reciprocating engines may be configured to carry as many as ____ passengers.
   a. 75
   b. 90
   c. 105
   d. 120

____ 19. Gas turbine engines generate power by the rapid ____ of the fuel/air mixture when ignited.
   a. Vaporization
   b. Expansion
   c. Heating
   d. Combustion

____ 20. Air is drawn into the compressor section of gas turbine engines where it is compressed and accelerated by ____.
   a. Convection chambers
   b. Expansion fins
   c. Barometric force
   d. Rotating blades

____ 21. In gas turbine engines, the high-speed gases cause the turbines to rotate, which in turn drives the ____.
   a. Compressor section
   b. Combustion chamber
   c. Drive shaft
   d. Fuel control unit

____ 22. In gas turbine engines, the fuel pump, hydraulic pump, oil pump, and cooler are contained in the ____.
   a. Accessory section
   b. Junction box
   c. Auxiliary unit
   d. Secondary motor

____ 23. The simplest of the gas turbine engines is the ____.
   a. Turboprop
   b. Turboshaft
   c. Turbojet
   d. Turbofan

____ 24. Which type of gas turbine engine is most commonly found on large jetliners?
   a. Turboprop
   b. Turboshaft
   c. Turbojet
   d. Turbofan

____ 25. Which type of gas turbine engine is most commonly used in helicopters?
   a. Turboprop
   b. Turboshaft
   c. Turbojet
   d. Turbofan

____ 26. Turboprop engines are easily distinguished from piston engines by the turboprop's _____.
   a. Number of propeller blades
   b. Smaller exhaust port
   c. More streamlined engine nacelle
   d. Quieter operation

____ 27. Which type of aircraft has variable pitch exhaust nozzles?
   a. F-16 fighter jet
   b. Harrier attack jet
   c. Sikorsky Sky Crane
   d. A-10 attack jet

____ 28. A gas turbine engine component that provides additional thrust for short periods is a(n) _____.
   a. Exhaust nozzle
   b. Turbocharger
   c. Afterburner
   d. Thrust-reversal system

____ 29. A gas turbine engine component that assists in slowing down the aircraft during its landing rollout is a(n) _____.
   a. Exhaust nozzle
   b. Turbocharger
   c. Afterburner
   d. Thrust-reversal system

____ 30. Flight controls, braking systems, landing gear bay doors, and cargo door operation rely heavily on _____.
   a. Hydraulic pressure
   b. Electric motors
   c. Pneumatic pressure
   d. Gears and drive belts

____ 31. When approaching a propeller, personnel should remain at least _____ from it.
   a. 10 feet *(3 m)*
   b. 15 feet *(5 m)*
   c. 20 feet *(6 m)*
   d. 25 feet *(8 m)*

____ 32. In gusty wind conditions, a helicopter's main rotor may dip to within _____ of the ground.
   a. 3 feet *(1 m)*
   b. 4 feet *(1.3 m)*
   c. 5 feet *(1.5 m)*
   d. 6 feet *(1.8 m)*

____ 33. When landing, helicopters must have sufficient clearance of all ground cover within _____.
   a. 200 feet *(60 m)*
   b. 150 feet *(46 m)*
   c. 100 feet *(33 m)*
   d. 50 feet *(15 m)*

____ 34. To avoid suction hazards from jet engines, personnel should stay at least _____ away from the front and sides of the engine.
   a. 30 feet *(10 m)*
   b. 40 feet *(12 m)*
   c. 50 feet *(15 m)*
   d. 60 feet *(18 m)*

3

___ 35. A medium-sized jet transport aircraft's ground-idle blast danger area extends ___ from the back of the aircraft.
   a. 450 feet *(137 m)*
   b. 500 feet *(152 m)*
   c. 600 feet *(183 m)*
   d. 1,200 feet *(366 m)*

___ 36. After shutdown, jet engines retain sufficient heat to ignite spilled flammable materials for up to ___.
   a. 5 minutes
   b. 10 minutes
   c. 15 minutes
   d. 20 minutes

___ 37. A green light can be found at the aircraft's ___.
   a. Left wingtip
   b. Tail section
   c. Right wingtip
   d. Vertical stabilizer

___ 38. Rotating or flashing red anti-collision lights are also used to indicate that ___.
   a. Aircraft are backing.
   b. Aircraft are in motion.
   c. Aircraft are taking off.
   d. Aircraft engines are operating.

___ 39. Which class of storage compartment has a separate, approved smoke or fire detection system and sufficient access in flight to enable a crew member to effectively reach any part of the compartment with a handheld fire extinguisher?
   a. Class A
   b. Class B
   c. Class C
   d. Class E

___ 40. Which class of cargo compartment is the entire cabin of an all-cargo airplane?
   a. Class A
   b. Class B
   c. Class C
   d. Class E

___ 41. In which class of cargo compartment would a fire be easily discovered by a crew member while at his or her station?
   a. Class A
   b. Class B
   c. Class C
   d. Class E

___ 42. Aluminum comprises ___ of aircraft construction.
   a. 70 percent
   b. 75 percent
   c. 80 percent
   d. 85 percent

___ 43. Which material is used where high strength is required, even though its weight per volume is much higher than other structural materials?
   a. Titanium
   b. Aluminum
   c. Steel
   d. Magnesium

____ 44. Which material is used for certain engine parts because it is both strong and lightweight?
   a. Titanium
   b. Aluminum
   c. Steel
   d. Magnesium

____ 45. Which material is used to reinforce skin surfaces to protect them from impinging exhaust flames or heat?
   a. Titanium
   b. Aluminum
   c. Steel
   d. Magnesium

____ 46. Which type of aircraft incorporates elaborate wood fixtures for interior furnishings?
   a. Commercial transport
   b. Cargo
   c. General aviation
   d. Corporate

____ 47. A color-coded label with a blue vertical stripe on the left, a yellow vertical stripe in the center, and a white vertical stripe with circles on the right designate a _____ component.
   a. Lubrication
   b. Hydraulic
   c. Instrument air
   d. Pneumatic

____ 48. A color-coded label whose left two-thirds are a brown band and whose right one-third is a white stripe with diamonds designates a _____ component.
   a. Fuel
   b. Breathing oxygen
   c. Fire protection
   d. Deicing

____ 49. In addition to the wing area, commercial aircraft store fuel in the _____.
   a. Cargo bay
   b. Center fuselage
   c. Vertical stabilizer
   d. Nacelles

____ 50. During flight operations, the fuel in auxiliary tanks is usually consumed _____.
   a. First
   b. Last
   c. When the pilot manually switches to those tanks
   d. Immediately before landing

____ 51. In gravity fueling, fuel tanks may be filled through _____.
   a. A single point on the underside of the wings
   b. Multiple points on the underside of the wings
   c. Service openings on the topside of the wings
   d. The side of the fuselage

3

_____ 52. The best way to shut down fuel pumps on an aircraft is to _____.
   a. Shut down the electrical system.
   b. Close valves manually at the fuel tank.
   c. Locate the fuel pump and manually turn it off.
   d. Secure power and fuel controls in the flight deck area.

_____ 53. To reduce pressure buildup caused by expansion of fuels, fuel tanks are equipped with _____.
   a. Overflow valves   c. Spill pans
   b. Siphon jets   d. Vent tanks

_____ 54. Most modern aircraft hydraulic systems operate at a pressure of _____ or higher.
   a. 2,000 psi *(14 000 kPa)*   c. 4,000 psi *(28 000 kPa)*
   b. 3,000 psi *(21 000 kPa)*   d. 5,000 psi *(34 000 kPa)*

_____ 55. The most widely used hydraulic fluids are _____.
   a. Water-based   c. Hydrocarbon-based
   b. Synthetic   d. Vegetable-based

_____ 56. All of the following are drawbacks associated with synthetic hydraulic fluids *except* _____.
   a. Low flash point
   b. Skin and eye irritation
   c. Extremely flammable in the form of a fine mist
   d. May ignite if sprayed on hot brakes

_____ 57. When dealing with a landing gear emergency, always approach the landing gear from _____.
   a. Either side   c. The aft only
   b. The aft or the side   d. Either forward or aft

_____ 58. Large-aircraft electrical systems operate on _____.
   a. 24/28-volt DC only
   b. 12- or 24-volt DC and 110/115-volt AC
   c. 24/28-volt DC and 110/115-volt AC
   d. 24/48-volt DC and 210/220-volt AC

_____ 59. Both lead-acid and nickel-cadmium batteries produce _____ when charging.
   a. Water vapor   c. Hydrogen gas
   b. Carbon dioxide gas   d. Ammonia gas

_____ 60. Most commercial and military aircraft batteries are equipped with _____ terminals.
   a. Permanent-connection   c. Fail-safe
   b. Quick-disconnect   d. Alligator-clip

30
Aircraft Rescue and Fire Fighting

61. While the aircraft is airborne, _____ can sometimes be used as a backup electrical power source.
    a. The battery
    b. One of the engines
    c. The CPU
    d. The APU

62. The APU generally is located in the _____ of the aircraft.
    a. Tail section
    b. Fuselage between the wings
    c. Wings
    d. Flight deck

63. Manual controls for the APU are located in the cockpit and _____.
    a. On the APU itself
    b. On an external fire-protection panel
    c. In the cargo compartment
    d. At the main flight attendant's station

64. Which type of emergency power unit may deploy when the electrical system is de-energized?
    a. Ram-air-turbine
    b. Jet-fuel
    c. Monopropellant
    d. Fuel-cell

65. Monopropellant emergency power units use a toxic and caustic fuel called _____.
    a. Benzene
    b. Formula 1
    c. Hydrazine
    d. Ammonium nitrogen

66. Fuels that ignite spontaneously on contact with an oxidizer are classified as _____.
    a. Anhydrous
    b. Hypergolic
    c. Spontaneous agents
    d. Explosive

67. In most cases, oxygen cylinders aboard aircraft are painted _____.
    a. Light blue
    b. Red
    c. Orange
    d. Green

68. Which of the following is *not* a hazard of an aircraft oxygen system?
    a. Fires will burn with more intensity in an oxygen-enriched environment.
    b. An explosion may result if liquid oxygen mixes with flammable/combustible materials.
    c. A deflagration may result if an oxygen storage tank ruptures.
    d. Imploded oxygen tanks may overheat and ignite nearby flammables.

____ 69. Most airborne radar systems are operated on the ground only before takeoff and just after landing because ____.
   a. Radar can interfere with radio communications.
   b. Radar energy can present an ignition source and a health hazard.
   c. Radar energy will deplete the aircraft's reserve energy supply.
   d. Radar is ineffective on the ground.

____ 70. Anti-icing units may be ____ and/or ____.
   a. Electric; pneumatic
   b. Hydraulic; pneumatic
   c. Electric; chemical propulsion
   d. Hydraulic; electrostatic

____ 71. Heated, L-shaped probes that protrude from the sides of the fuselage are ____.
   a. Radar antennae   c. Pitot tubes
   b. Temperature gauges   d. Radio transmitters

____ 72. A common first step in shutting down any aircraft is to ____.
   a. Move the throttle(s) to the IDLE or OFF position.
   b. Shut off the battery switches.
   c. Activate the fire-protection system.
   d. Activate the engine and APU fire shutoff handles.

____ 73. The T- or L-shaped engine and APU fire shutoff handles are usually located ____.
   a. On the instrument panel   c. In the battery compartment
   b. At the exit(s)   d. Around the throttles

____ 74. Aircraft generally are designed to be evacuated in ____ or less in the event of an emergency.
   a. 60 seconds   c. 120 seconds
   b. 90 seconds   d. 150 seconds

____ 75. The door indicated in the figure below would be designated ____.

   a. R2   c. L2
   b. R3   d. L3

3

_____ 76. Which of the following statements about aircraft doors is true?
  a. Opening and operating procedures will be the same for all doors found on the same aircraft.
  b. Primary egress from aircraft is through emergency doors and escape hatches.
  c. All primary access doors on aircraft have an exterior latch release that disconnects the locking device and permits the door to open.
  d. Types of cabin doors are limited to four standard variations.

_____ 77. Which type of aircraft may have doors that open by moving upward into the fuselage?
  a. Business/commuter
  b. Narrow-bodied commercial
  c. General aviation
  d. Wide-bodied commercial

_____ 78. What color floor lighting in commercial airliners indicates an emergency exit?
  a. Amber
  b. Green
  c. Red
  d. White

_____ 79. The FAA requires that an aircraft be equipped with an inflatable emergency escape slide if the bottom doorsill is greater than _____ from the ground.
  a. 6 feet *(2 m)*
  b. 8 feet *(2.4 m)*
  c. 10 feet *(3 m)*
  d. 12 feet *(3.7 m)*

_____ 80. If ground ladders must be used to open escape-slide doors, they should be placed _____.
  a. Under the door
  b. In front of the door
  c. Beside the door on the side opposite the hinges
  d. Beside the door on the same side as the hinges

_____ 81. As many as _____ of aircraft occupants suffer minor to moderate injuries going down escape slides.
  a. 5 to 10 percent
  b. 10 to 15 percent
  c. 15 to 20 percent
  d. 20 to 25 percent

_____ 82. One thing that escape hatches or windows on pressurized aircraft have in common is that they _____.
  a. Are plug type
  b. Fall away from the aircraft
  c. Are equipped with escape slides
  d. Must be pulled inward from the inside

_____ 83. The Boeing® "automatic over-wing exit door" is operated by a _____ mechanism.
  a. Hand-powered
  b. Powder-activated
  c. Electrically powered
  d. Spring-loaded

33

Aircraft Rescue and Fire Fighting

84. Before using stairs built into the rear of an aircraft as a means of egress, firefighters should ensure that the _____.
    a. Stairs have an adequate weight rating
    b. Aircraft is stabilized
    c. Electrical system has been shut down
    d. Forward exits cannot be opened

85. The tail-cone jettison system requires passengers to exit by _____.
    a. Jumping
    b. Climbing down a rope ladder
    c. Walking down stairs
    d. Using an escape slide

86. Before attempting to open the main cabin doors of an aircraft with a pressurized cabin, firefighters should find and force open the _____.
    a. Over-/under-wing escape hatches
    b. Tail-cone jettison system
    c. Outflow valve
    d. Exterior APU control panel hatch

87. Firefighters should initiate emergency cut-in procedures _____.
    a. As soon as the appropriate resources arrive
    b. Immediately upon determining that the outflow valve is closed
    c. Whenever a primary means of egress is unavailable
    d. Only as a last resort

88. What should be done with a data-recording unit that is found submerged in water if there is a chance of its being lost?
    a. It should be anchored and marked with a buoy.
    b. It should be removed and stored in an airtight container.
    c. It should be removed and stored in fresh water.
    d. Qualified personnel should be posted to watch it.

89. In military aircraft designations, the letter *E* stands for _____.
    a. Emergency services
    b. Special electronic installation
    c. Extended duty
    d. Medical evacuation

90. In military aircraft designations, the letter *S* stands for _____.
    a. Antisubmarine
    b. Strategic observation
    c. Air surveillance
    d. Special electronic installation

91. Military aircraft designed to engage in air-to-air combat and/or attack targets on the ground are called _____.
    a. Fighter aircraft
    b. Assault aircraft
    c. Light bomber aircraft
    d. Raid aircraft

92. Military aircraft designed to carry and drop a large quantity of air-to-ground weapons are called _____.
    a. Attack aircraft
    b. Heavy ordnance aircraft
    c. Strategic command aircraft
    d. Bomber aircraft

93. Cargo aircraft may have _____.
    a. Ejection seats
    b. Canopy removal systems
    c. Jet-assisted takeoff units
    d. Weapon racks and externally mounted fuel tanks

94. Tanker aircraft may carry fuel loads of up to _____.
    a. 25,000 gallons *(95 000 L)*
    b. 50,000 gallons *(200 000 L)*
    c. 75,000 gallons *(285 000 L)*
    d. 100,000 gallons *(380 000 L)*

95. Which type of military aircraft is generally quite similar to general aviation aircraft?
    a. Cargo
    b. Utility
    c. All-purpose
    d. Staff

96. Weapons attached to military helicopters are usually carried inside the cabin or _____.
    a. Under the nose
    b. On the tail
    c. On pods attached to the fuselage
    d. On the landing skids

97. Almost all military aircraft have either _____ or _____ fire extinguishing systems to protect the engines.
    a. Halon; nitrogen
    b. Water; dry powder
    c. AFFF; nitrogen
    d. Dry powder; AFFF

98. Ejection systems that may be fired while the aircraft is on the ground and parked are referred to as _____ systems.
    a. Ground-ground
    b. Go-go
    c. Start-finish
    d. Zero-zero

99. Which type of canopy is shown in the illustration below?

    a. Sliding
    b. Clamshell
    c. Hinge
    d. Pop-off

3

____ 100. Canopy ejection systems are fired by _____ devices.
    a. Pneumatic
    b. Hydraulic
    c. Spring-loaded
    d. Explosive

____ 101. Canopy removers, initiators, rotary actuators, explosive squibs, and seat catapults are examples of _____.
    a. Pressure-triggering mechanisms
    b. Power-plus activators
    c. Propellant actuating devices
    d. Pilot-command levers

____ 102. Generally, canopy removers are _____ devices.
    a. Hydraulic
    b. Gas-pressured
    c. Spring-loaded
    d. Pneumatic

____ 103. Cylindrically shaped devices that provide the gas pressure to start a sequence of events in the emergency ejection process are called _____.
    a. Initiators
    b. Activators
    c. Starters
    d. Pressure triggers

____ 104. Which type of propellant actuating device separates the crew member from the seat after ejection?
    a. Thruster
    b. Explosive squib
    c. Rotary actuator
    d. Seat catapult

____ 105. Which type of propellant actuating device unlocks the canopy latches just before canopy jettison?
    a. Thruster
    b. Explosive squib
    c. Rotary actuator
    d. Seat catapult

____ 106. A small metal tube that contains a flammable mixture that produces pressure or provides an ignition source is a(n) _____.
    a. Thruster
    b. Explosive squib
    c. Rotary actuator
    d. Seat catapult

____ 107. Telescoping devices used in the emergency ejection of the aircrew are _____.
    a. Thrusters
    b. Explosive squibs
    c. Rotary actuators
    d. Seat catapults

____ 108. Which of the following statements about pressed high explosives is true?
    a. In a fire situation where the ammunition case is not ruptured, detonation is unlikely.
    b. If the ammunition case is broken open and the explosive begins to burn, it will lose its explosive properties.
    c. Burning pressed high explosives may produce flames of almost any color.
    d. Pressed high explosives generally produce dull, smoky flames with low light intensity.

3

____ 109. When approaching a fighter or attack aircraft, personnel should take position at a ____ angle off the nose or tail, provided it does not place them in front of or behind under-wing rockets or missiles.
   a. 30-degree
   b. 45-degree
   c. 60-degree
   d. 90-degree

____ 110. Photoflash cartridges and high-intensity flares are types of ____.
   a. Pyrotechnics
   b. Mono-propellants
   c. High explosives
   d. Antioxidizers

____ 111. The difference between rockets and missiles is that ____.
   a. Rockets are larger than missiles.
   b. Missiles carry warheads and rockets do not.
   c. Missiles have a guidance-and-control system and rockets do not.
   d. Rockets are self-propelled and missiles are not.

____ 112. When a gravity bomb is involved in fire and cannot be cooled quickly, the area must be evacuated to no less than ____.
   a. 500 feet *(152 m)*
   b. 1,000 feet *(300 m)*
   c. 1,500 feet *(455 m)*
   d. 2,000 feet *(610 m)*

____ 113. Depending on the type of weapon involved in a fire, conventional weapons or explosives may be expected to detonate in as little as ____.
   a. 10 seconds
   b. 30 seconds
   c. 45 seconds
   d. 60 seconds

Identify

D. **Identify the following abbreviations associated with aircraft. Write the correct interpretation before each.**

_____ 1. NLA

_____ 2. VLA

_____ 3. AVGAS

_____ 4. EPU

_____ 5. RAT

_____ 6. LOX

_____ 7. FDR

_____ 8. CVR

_____ 9. NTSB

3

_____ 10. JATO

_____ 11. HE

_____ 12. EOD

E. **Identify procedures for helicopter safety. Write an *X* before each correct statement below.**

____ 1. Wait until the pilot signals that it is safe to approach the aircraft.

____ 2. Approach a helicopter from the rear.

____ 3. Approach a helicopter from the uphill side.

____ 4. Carry tools horizontally, below waist level.

____ 5. When directing a helicopter for landing, stand facing the wind.

Chapter 4 ARFF Firefighter Safety

True/False

A. Write *True* or *False* before each of the following statements. Correct those statements that are false.

_____ 1. Each firefighter not only is responsible for his or her own safety but also must look out for the safety of the entire team.

_____ 2. Simply wearing a PASS device increases a firefighter's chances of being found in an emergency.

_____ 3. The use of hearing protection is important in and around the fire station as well as during fire-fighting operations.

_____ 4. Firefighters are solely responsible for knowing and obeying the rules for their own vision protection.

_____ 5. Helmet-mounted face shields provide adequate vision protection for all fire-fighting activities.

_____ 6. Firefighters who must respond to aircraft emergencies with only structural fire-fighting protective equipment are adequately protected from all but the most extreme conditions.

4

_____ 7. With the addition of one or more layers of thermal barrier, proximity suits can withstand exposure to steam, liquids, and some weaker chemicals.

_____ 8. The two-in/two-out concept is no different for aircraft fires than for structural fires.

_____ 9. Personnel decontamination generally is not needed at aircraft crash sites unless the aircraft was carrying hazardous cargo.

_____ 10. One of the greatest silent killers in any emergency responder's life is stress.

Multiple Choice

B. Write the letter of the best answer on the blank before each statement.

____ 1. The *Standard on Fire Department Occupational Safety and Health Program* is ____.
  a. NFPA 1000
  b. NFPA 1500
  c. NFPA 2000
  d. NFPA 1500

____ 2. Many civilian aircraft use carbon and other graphite fibers in construction, creating a hazard similar to ____.
  a. Asbestos
  b. Carbon monoxide
  c. Silicon
  d. Coal dust

____ 3. The *Standard on Personal Alert Safety Systems (PASS) for Fire Fighters* is ____.
  a. NFPA 1582
  b. NFPA 1825
  c. NFPA 1892
  d. NFPA 1982

40
Aircraft Rescue and Fire Fighting

_____ 4. A PASS device should be capable of emitting an alarm of _____ decibels.
   a. 75
   b. 85
   c. 95
   d. 105

_____ 5. The most common problem with PASS devices is _____.
   a. Smoke damage
   b. Dead batteries
   c. Heat damage
   d. Age-related failure

_____ 6. How have some manufacturers removed the problem of firefighters forgetting to activate PASS devices?
   a. By making a device that is always on
   b. By integrating the device into the SCBA
   c. By printing reminders on the device
   d. By placing a switch in the device's attachment mechanism

_____ 7. Which NFPA standard defines the maximum level of noise to which fire protection personnel are allowed to be exposed in the work environment?
   a. NFPA 1001
   b. NFPA 1582
   c. NFPA 1500
   d. NFPA 1982

_____ 8. The bottom line with a hearing awareness program is to _____.
   a. Wear hearing protection.
   b. Have regular hearing tests.
   c. Avoid exposure to loud noises.
   d. Attend regular refresher courses.

_____ 9. Firefighters assigned to an airport response should use _____.
   a. Haz mat suits
   b. Structural fire-fighting protective clothing
   c. Approach clothing
   d. Proximity suits

_____ 10. The U.S. Environmental Protection Agency classifies work uniforms as _____ protection, suitable only for routine support functions.
   a. Level 1
   b. Level 5
   c. Level A
   d. Level D

_____ 11. The main advantage of proximity suits over structural fire-fighting protective clothing is that proximity suits _____.
   a. Resist cuts and abrasions
   b. Reflect radiant heat
   c. Have a moisture barrier
   d. Provide protection against hazardous materials

4

_____ 12. For information on proper haz mat protective clothing levels, firefighters should refer to _____.
   a. OSHA 32.A.1
   b. EPA Title 12
   c. NFPA 471
   d. DOT *Emergency Response Guidebook*

_____ 13. Aircraft accidents require a well-organized and well-trained emergency crew using the _____.
   a. Strategic Planning Model
   b. Incident Management System
   c. Emergency Command Structure
   d. Tactical Response Organization

_____ 14. Which of the following statements about ARFF personnel accountability is true?
   a. Having a sound accountability system makes for a well-organized emergency.
   b. Accounting for all personnel typically is less difficult during aircraft emergencies than during structural fire fighting.
   c. Even if a department follows good, standard IMS rules, accounting for its firefighters will ordinarily present major problems.
   d. An ARFF department's first step to account for its personnel is to write a comprehensive standard operating procedure.

_____ 15. U.S. regulations allow an exception to the two-in/two-out rule if _____.
   a. A victim is known to be accessible to rescuers.
   b. A known victim cannot be located.
   c. The needed additional firefighters are known to be on their way to the scene.
   d. Qualified civilians are on hand.

_____ 16. How often should ARFF personnel practice following their department's two-in/two-out policy?
   a. Every six months
   b. Annually
   c. Often enough to maintain a basic familiarity with it
   d. Often enough for it to become second nature

_____ 17. The main reason why stress affects physical and emotional health is that _____.
   a. Individual weakness creates a need to escape.
   b. Physical exertion increases negative stressors.
   c. The mind and body work against each other.
   d. It interferes with normal thought processes.

____ 18. Which of the following statements about critical incident stress debriefing is true?
   a. Participation in CISD should be optional for all individuals involved in mass-casualty incidents.
   b. Personnel should always seek CISD immediately after any critical situation is resolved.
   c. Firefighters' knowledge that they have all their skills and specialized equipment can help to reduce the trauma of responding to a large passenger aircraft crash when there are no survivors.
   d. Firefighters should be told what to expect before they enter a mass-casualty scene.

____ 19. If firefighters are required to work more than one shift under critical incident stress conditions, how soon after completing their work at the incident should they participate in the full debriefing process?
   a. Within 24 hours
   b. Within 48 hours
   c. Within 72 hours
   d. Within 7 days

____ 20. Which of the following statements about firefighter safety at the fire station is true?
   a. It is the department's safety officer's responsibility to look out for all the safety hazards in the fire station.
   b. Personnel who observe any situations that warrant a safety concern should bring them to the attention of the health and safety officer.
   c. Material safety data sheets should be kept in a secure place in a safe location away from the fire station.
   d. Portable heaters should be clearly marked and placed in travel routes.

Identify

C. **Identify the following abbreviations associated with ARFF firefighter safety. Write the correct interpretation before each.**

_____ 1. PASS

_____ 2. IMS

_____ 3. CISD

_____ 4. MSDS

D. Identify hazards associated with ARFF. Write an *X* before each correct statement below.

___ 1. Aircraft jet engines can ingest firefighters and overturn vehicles with the jet wash.

___ 2. Reciprocating engines cannot be restarted simply by moving the prop.

___ 3. Landing gears are made from special materials that prevent violent reactions when water or foam is applied.

___ 4. Jet fuel is a known carcinogen.

___ 5. Aircraft electrical systems do not generate enough power to electrocute personnel.

___ 6. Hydraulic and pneumatic lines contain flammable and toxic fluids and gases.

___ 7. All pressurized oxygen systems pose a significant risk of explosion when engulfed in flames.

___ 8. EPA and OSHA regulations prohibit the use of composite materials that contain potentially hazardous fibers in aircraft construction.

___ 9. Biohazards at aircraft crash sites come mainly from the bodily fluids of occupants of the aircraft.

___ 10. Depleted uranium used for counterweights and energized radar systems can be threats to emergency responders.

Chapter 5 Aircraft Rescue and Fire Fighting Communications

True/False

A. Write *True* or *False* before each of the following statements. Correct those statements that are false.

_____ 1. Because incident communications are conducted internally, they do not affect the fire department's public image.

_____
_____

_____ 2. Ideally, ARFF communications should be coordinated with other agencies within the area.

_____
_____

_____ 3. FAA requirements mandate the use of a nationally standardized system to notify firefighters of an aircraft incident/accident.

_____
_____

_____ 4. Response categories for airport incidents are nationally standardized in the United States and Canada.

_____
_____

_____ 5. When airport auxiliary firefighters are used, they may be notified by pagers.

_____
_____

_____ 6. The common traffic advisory frequency is used on airports without an operating ATCT or when the tower is closed.

_____
_____

5

_____ 7. When ARFF personnel must communicate with ground control, they should repeat the tower instructions before acting.

_____ 8. During ARFF emergency operations, the incident commander directs aircraft and vehicular traffic by two-way radio on the ground-control frequency.

_____ 9. Airport control towers use light signals in conjunction with radio communications.

Multiple Choice

B. Write the letter of the best answer on the blank before each statement.

_____ 1. The suggested methodology for planning and implementing ARFF communications can be found in _____.
   a. NFPA 1500
   b. FAA Advisory Circular 150/5210-7C
   c. FCC Regulation IV.G.7
   d. NFPA 471

_____ 2. If fire protection apparatus and services fall below the requirements identified in FAR Part 139.319 for longer than _____, fire department personnel must notify the airport operator so that a NOTAM can be issued.
   a. 1 hour   c. 24 hours
   b. 12 hours   d. 48 hours

_____ 3. Airport communications for ARFF operations include means for all of the following *except* _____.
   a. Notifying local news media   c. Using direct-line telephones
   b. Sounding audible alarms   d. Using radios

_____ 4. In ARFF communications, the amount of fuel an aircraft has on board is usually given in _____.
   a. Gallons   c. Pounds
   b. Hours of flying time remaining   d. Cubic feet

5. Which of the following statements about direct-line telephone communications is true?
   a. Direct-line communication is limited to that between the control tower and the ARFF station.
   b. All direct-line telephone communications systems allow two-way conversation.
   c. Direct-line systems are limited to notifying and requesting resources from only one organization at a time.
   d. Direct-line telephone conference circuits may include airline station managers, medical transport organizations, area hospitals, and mutual aid fire departments.

6. The most efficient means for communicating with personnel on the emergency scene operations is _____.
   a. Face-to-face meetings
   b. Two-way radio
   c. Cellular telephones
   d. Direct-line telephones

7. In order to monitor local radio channels for critical emergency information, agencies should have _____.
   a. Multichannel scanning capability
   b. As many radios as there are agencies involved
   c. A direct-line telephone system
   d. A common set of priority codes

8. During multiagency operations, the use of _____ helps eliminate confusion.
   a. 10-codes
   b. Clear Text
   c. Local codes and terminology
   d. Tech Speech

9. The document entitled "Rules Governing Public Safety Radio Service" is _____.
   a. NFPA 95
   b. FAR IV.D
   c. ICAO C-1
   d. FCC Part 89

10. The communications/dispatch center's duties include all of the following *except* _____.
    a. Clearing the air as soon as possible
    b. Maintaining discipline on the air
    c. Determining the order of priority for simultaneous transmissions
    d. Ensuring an open channel for news media communications

11. To obtain clearance for driving on the aircraft movement area at a controlled airport, firefighters will use a(n) _____.
    a. Regional control frequency
    b. Ground control radio frequency
    c. Local control frequency
    d. Emergency response frequency

12. From the time they are transferred to the airport air traffic control tower until they turn off the runway onto a taxiway, aircraft are on the ____.
    a. Regional control frequency
    b. Ground control radio frequency
    c. Local control frequency
    d. Emergency response frequency

13. At airports without an active tower or airport advisory, a radio frequency may be operated by ____.
    a. Flight Service Stations
    b. Ham radio operators
    c. Aviation clubs
    d. Aeronautic Communications Cooperatives

14. A private, nongovernmental frequency that may provide information or access to services and is usually found at general aviation airports is ____.
    a. VIA VOICE
    b. TransComm
    c. UNICOM
    d. Vox Populi

15. Which frequency has a continuous radio broadcast on weather and airfield information?
    a. CTAF
    b. FSS
    c. UNICOM
    d. ATIS

16. Who should initiate pilot/ARFF IC communications during an aircraft emergency?
    a. Control tower
    b. Flight engineer
    c. Pilot
    d. ARFF IC

17. Guidance for initiating the pilot/ARFF communications procedure can be found in ____.
    a. NFPA 1500
    b. FAA AC 150/5210-7C
    c. FCC Regulation 1998.92A
    d. NOTAM 11.17.1996

18. The interphone system that allows communication with the flight deck as well as various compartments, wheel wells, rear empennage access areas, fueling and APU panels, and other areas on the aircraft is the ____.
    a. Flight connection
    b. All-purpose connection
    c. Duty connection
    d. Service connection

19. The ARFF hand signal in which a firefighter extends one arm horizontally from the body with the hand upraised at eye level and makes a beckoning motion indicates ____.
    a. Recommended evacuation
    b. Clear to land
    c. Move the aircraft forward
    d. Proceed to taxiway

5

____ 20. The ARFF hand signal in which a firefighter stands with arms in front of the head and crossed at the wrists indicates _____.
   a. Move the aircraft backward
   b. Recommended evacuation
   c. Clear the runway/taxiway
   d. Recommended stop

____ 21. The ARFF hand signal that indicates "emergency contained" is the _____.
   a. Football "touchdown" signal
   b. Baseball "safe" signal
   c. Baseball "strike" signal
   d. Basketball "basket counts" signal

Identify

C. Identify the following abbreviations associated with ARFF communications. Write the correct interpretation before each.

_____ 1. NOTAM

_____ 2. ATCT

_____ 3. FSS

_____ 4. VFR

_____ 5. UNICOM

_____ 6. CTAF

_____ 7. ATIS

_____ 8. MDT

_____ 9. GPS

D. Identify the correct order of information for communicating on the ground control frequency. Write each step's sequential number in the blank provided.

   ____ A. Vehicle identity such as "ARFF 1"

   ____ B. Preferred route to take

   ____ C. Request of clearance to desired area

   ____ D. Name of facility being called

   ____ E. Firefighter location

49

Aircraft Rescue and Fire Fighting

5

E. **Identify guidelines for proper radio/telephone use. Write an *X* before each correct statement below.**

____ 1. Hold the microphone parallel to the plane of the face, at least 2½ inches *(64 mm)* from the mouth.

____ 2. Convey messages word by word rather than in phrases.

____ 3. Use an urgent tone.

____ 4. Speak only as loudly as you would in ordinary conversation. If surrounding noise interferes, speak louder, but do not shout.

____ 5. Try to speak in a high-pitched voice because high-pitched tones transmit better than low-pitched tones.

F. **Identify letters of the ICAO phonetic alphabet. Write the standard word for each letter in the blank provided.**

_____ A   _____ N

_____ B   _____ O

_____ C   _____ P

_____ D   _____ Q

_____ E   _____ R

_____ F   _____ S

_____ G   _____ T

_____ H   _____ U

_____ I   _____ V

_____ J   _____ W

_____ K   _____ X

_____ L   _____ Y

_____ M   _____ Z

5

G. **Identify words and phrases in most common usage in the airport environment. Write the correct word or phrase for each definition in the space provided.**

_____ 1. Confirm that you have received and understood the message.

_____ 2. An error has been made in the transmission, and the corrected version follows.

_____ 3. Unintended loss of combustion in turbojet engines resulting in the loss of engine power.

_____ 4. Do not proceed! Remain where you are.

_____ 5. An approach over a runway or heliport where the pilot intentionally does not make contact with the runway.

_____ 6. The international radio distress signal.

_____ 7. The conversation is ended, and no response is expected.

_____ 8. My transmission is ended; I expect a response.

_____ 9. Message received and understood.

_____ 10. Prepare to receive detailed information that should be written down.

_____ 11. Request for confirmation of information.

_____ 12. Received message, understand, and will comply.

H. **Identify light-gun signals. Write the correct meaning for each signal in the space provided.**

1. Steady green light _____

2. Steady red light _____

3. Flashing red light _____

4. Flashing white light _____

5. Alternating red and green lights _____

I. **Identify other signals for aircraft accident operations. Write the correct signal in the space provided.**

1. Back out or retreat _____
   _____

2. Apparatus is running out of agent _____
   _____

5

3. Open or close handline _____

_____

4. Change handline nozzle/stream pattern _____

_____

5. Back out with handline _____

_____

Chapter 6 Aircraft Rescue and Fire Fighting Apparatus

True/False

A. Write *True* or *False* before each of the following statements. Correct those statements that are false.

_____ 1. Aviation industry standards require that ARFF apparatus be able to reach the scene of an aircraft emergency in the same amount of time as municipal departments are able to respond to a structure fire.

_____

_____

_____ 2. Apparatus with fixed foam proportioning systems give firefighters more capability than an external inline foam eductor.

_____

_____

_____ 3. Batch mixing is suitable for both regular and alcohol-resistant AFFF concentrates.

_____

_____

_____ 4. Modified mobile air stairs, food service vehicles, and similar devices generally count toward the minimum index requirements for fire protection.

_____

_____

_____ 5. Antilock brakes can give a driver a false sense of security if the driver completely relies on the braking system to keep the vehicle under control.

_____

_____

_____ 6. The DEVS includes an infrared camera and monitor that enhance vision in smoke, fog, adverse weather, and darkness.

_____

_____

6

_____ 7. Handlines on ARFF apparatus must be equipped with nonaspirating nozzles.

_____ 8. Extendable turrets lower the center of gravity on ARFF vehicles.

_____ 9. Most major ARFF vehicles carry some type of auxiliary agent, which may include dry chemical or halon.

_____ 10. All ARFF vehicles should be able to quickly resupply with both water and foam concentrate.

_____ 11. All ARFF vehicles have overhead-fill method capabilities.

_____ 12. The least desired method of resupplying an apparatus with foam is to use a hand pump to transfer foam concentrate from large storage containers or a foam tender.

Multiple Choice

B. Write the letter of the best answer on the blank before each statement.

____ 1. At certificated airports in the U.S., airport management must notify the FAA and affected carriers of the reduction in operational readiness if any required apparatus or equipment should become inoperative and an equal replacement is not available within ____.
   a. 12 hours
   b. 24 hours
   c. 48 hours
   d. 72 hours

____ 2. Military aircraft operations are governed by ____.
   a. NFPA 403
   b. FAR Part 139.317
   c. ICAO *Airport Services Manual Part I*
   d. NFPA 412

____ 3. The *Standard for Aircraft Rescue and Fire Fighting Vehicles* is ____.
   a. NFPA 403   c. NFPA 412
   b. NFPA 405   d. NFPA 414

____ 4. An airport serving aircraft with an overall length up to but not including ____ and a width up to but not including ____ would have an NFPA rating of 6.
   a. 90 feet *(28 m)*; 13.0 feet *(4 m)*
   b. 126 feet *(39 m)*; 16.4 feet *(5 m)*
   c. 160 feet *(49 m)*; 16.4 feet *(5 m)*
   d. 200 feet *(61 m)*; 23.0 feet *(7 m)*

____ 5. An airport serving aircraft with an overall length up to but not including 59 feet *(18 m)* and a width up to but not including 9.8 feet *(3 m)* would have an FAA rating of ____.
   a. GA-1   c. A
   b. GA-2   d. B

____ 6. Which of the following statements about ARFF apparatus is true?
   a. ARFF apparatus are designed for on-road maneuvering only.
   b. ARFF vehicles should discharge extinguishing agents only when they are stationary.
   c. ARFF vehicles may have to make mass application of extinguishing agents to protect aircraft occupants.
   d. ARFF vehicles are equipped with the same equipment as structural firefighting apparatus.

____ 7. NFPA 414 classifies ARFF apparatus according to ____.
   a. Length   c. Number of personnel
   b. Intended use   d. Water-tank capacity

____ 8. The FAA would categorize an ARFF vehicle with a minimum usable water capacity of 2,500 gallons *(9 463 L)* as ____.
   a. Class 1   c. Class 3
   b. Class 2   d. Class 4

____ 9. The method of producing foam by pouring foam concentrate directly into the apparatus water tank is commonly referred to as ____.
   a. Batch mixing   c. Bulk suspension
   b. Direct blending   d. Low-concentrate storage

6

_____ 10. Which of the following statements about central inflation/deflation tire systems is true?
   a. Central inflation/deflation tire systems must be operated while the vehicle is stationary.
   b. Inflating the tire aids in removing mud and debris from the tread.
   c. Central inflation/deflation tire systems typically decrease downtime for vehicle maintenance.
   d. Deflating modern vehicle tires makes their aggressive tread more effective.

_____ 11. The DEVS component that provides a digital radio data link between the command center and vehicles is the _____ subsystem.
   a. Monitoring
   b. Tracking
   c. Communications
   d. Broad-band

_____ 12. The idea behind a high-mobility suspension system is to _____.
   a. Maximize ground clearance.
   b. Transfer enough power from the chassis to the surface to move obstacles.
   c. Keep the wheels in contact with the surface.
   d. Allow sharp turns at high speeds.

_____ 13. All of the following statements about fire pumps on ARFF apparatus are true *except* _____.
   a. All ARFF vehicles are capable of delivering large quantities of water to the fire-fighting systems.
   b. Fire pumps in ARFF vehicles can operate while the vehicle is in motion.
   c. Most manufacturers who provide ARFF apparatus with structural capability usually provide foam to all discharges.
   d. Structural fire-fighting capabilities impose serious disadvantages on ARFF vehicles.

_____ 14. The three types of turret nozzles used presently include all of the following *except* _____.
   a. Aspirating
   b. Nonaspirating
   c. Liquefying
   d. Dry-chemical injection

_____ 15. Which type of nozzle is used to lay a blanket or path of foam in front of an ARFF vehicle?
   a. Turret
   b. Extendable turret
   c. Ground-sweep nozzle
   d. Undertruck nozzle

_____ 16. Which type of nozzle discharges foam directly beneath the vehicle chassis?
   a. Turret
   b. Extendable turret
   c. Ground-sweep nozzle
   d. Undertruck nozzle

___ 17. The _____ is designed to attack the fire at the base of the flames.
   a. Turret
   b. Extendable turret
   c. Ground-sweep nozzle
   d. Undertruck nozzle

___ 18. The piercing nozzles on extendable turrets can flow in excess of _____.
   a. 200 gpm *(757 L/min)*
   b. 250 gpm *(946 L/min)*
   c. 350 gpm *(1 325 L/min)*
   d. 500 gpm *(1 893 L/min)*

___ 19. According to NFPA 414, the piping on ARFF apparatus must be designed so that filling the tank at an inlet pressure of 80 psi *(552 kPa)* requires _____ or less.
   a. 60 seconds
   b. 90 seconds
   c. 120 seconds
   d. 150 seconds

___ 20. The most effective method of supplying water to ARFF apparatus tank inlets is usually a supply line extended from the _____.
   a. Pumper
   b. Fire hydrant
   c. Mobile water supply
   d. Fixed, overhead-fill hose

___ 21. Which of the following statements about auxiliary-agent-system servicing is true?
   a. The most important thing for ARFF personnel to know when refilling is the type of agent their department uses.
   b. Certain types of agents or expellants may be mixed according to departmental guidelines.
   c. Dry chemical systems are commonly serviced during the course of an incident.
   d. Dry chemical may be left in the piping and hoselines.

___ 22. How often should ARFF apparatus be inspected and serviced?
   a. Immediately after shift change if it has been used during the previous shift; otherwise weekly
   b. Once during every shift
   c. Immediately after shift change and after each use
   d. Once a week and after each use

___ 23. Each pumper's permanent file should include a record of _____ pump performance tests.
   a. Monthly
   b. Quarterly
   c. Semiannual
   d. Annual

6

Identify

C. Identify the following abbreviations associated with ARFF apparatus. Write the correct interpretation before each.

_____ 1. DEVS

_____ 2. DGPS

_____ 3. FLIR

Chapter 7 Aircraft Rescue Tools and Equipment

True/False

A. Write *True* or *False* before each of the following statements. Correct those statements that are false.

_____ 1. In most cases, conventional fire-fighting tools are *not* adaptable to aircraft rescue and forcible entry.

_____
_____

_____ 2. Some power tools generate over 20,000 psi *(140 000 kPa)* of mechanical energy.

_____
_____

_____ 3. Any undercarriage fire creates a potential for aircraft collapse or the explosive disintegration of affected components.

_____
_____

_____ 4. To stabilize an aircraft, dirt can be pushed up against the fuselage or heavy equipment can be parked against it.

_____
_____

_____ 5. Techniques used for aircraft extrication are essentially the same as those used for auto extrication.

_____
_____

_____ 6. *ICAO Airport Services Manual, Part 1,* specifies the types and numbers of tools to be carried on ARFF vehicles.

_____
_____

7

_____ 7. Pneumo-hydraulic tools are not widely used because they are especially dangerous in flammable atmospheres.

_____ 8. Exceeding a battery-powered/electric drill/driver's recommended revolutions per minute may damage the opening mechanisms on aircraft compartments.

_____ 9. The pressure that drives hydraulically powered tools must be produced through a powered unit such as a gas-driven engine or electrical pump.

_____ 10. A disadvantage of using hydraulic tools in flammable areas is that they produce sparks.

_____ 11. A truck-mounted winch may be used to stabilize an aircraft or component.

_____ 12. The applications of rope in ARFF environments are more limited than in other fire-fighting operations.

_____ 13. An 8- to 10-foot *(2.4 m to 3 m)* length of webbing can be carried in the pocket of turnout gear.

_____ 14. Repair plugs may be made of wood or rubber.

_____ 15. Personnel should consider laddering the leading edge of the wing, all doors, and other aircraft access points.

_____ 16. If a modern, electrically operated aircraft door system fails, the only way to open the door is through forcible entry.

_____ 17. It may be possible to force open jammed doors on only the lightest and smallest of aircraft by using a pry bar around the frame or at the hinges.

_____ 18. Cuts should sever as many structural reinforcement members as practical.

Multiple Choice

B. Write the letter of the best answer on the blank before each statement.

____ 1. The four groups of aircraft rescue hand tools include all of the following *except* _____.

   a. Cutting
   b. Prying
   c. Pushing/pulling
   d. Extracting

____ 2. Spilled aircraft fuel can be covered by all of the following *except* _____.

   a. A foam blanket
   b. Water
   c. Dirt
   d. Absorbent materials

____ 3. Which of the following statements regarding safety in the operational area is true?

   a. New advanced aerospace materials eliminate the need for breathing apparatus.
   b. A high worker-to-supervisor/safety officer span of control should be maintained.
   c. Personnel should maintain a natural body position when using tools and equipment.
   d. As many firefighters as possible should remain in the operational area.

7

___ 4. Which NFPA standard specifies the types and numbers of tools to be carried on ARFF vehicles?
   a. NFPA 403
   b. NFPA 405
   c. NFPA 412
   d. NFPA 414

___ 5. Hand tools generally can be defined as tools that ____.
   a. Can be carried comfortably with one hand by an average-sized person
   b. Must be operated directly by human hands rather than robotics
   c. Rely on human force to transmit power directly to the working end of the tool
   d. Weigh 15 pounds (7 kg) or less and can be operated by only one person

___ 6. Which type of tool can be used to open access panels secured with Dzus fasteners?
   a. Screwdriver
   b. Pike pole
   c. Rescue tool assembly
   d. Axe

___ 7. Which type of tool would most likely be used to make holes for inserting the tips of hydraulic spreaders during forcible entry?
   a. Pike pole
   b. De-arming tool
   c. Axe
   d. Rescue tool assembly

___ 8. The tools shown in the figure below are two types of ____.

   a. Prying tools
   b. De-arming tools
   c. Metal shears
   d. Harness-cutting knives

___ 9. The tool shown in the figure below is a ____.

   a. Plier tool
   b. Dzus driver
   c. De-arming tool
   d. Cable cutter

Aircraft Rescue and Fire Fighting

_____ 10. Which power tools have pumps that produce and transmit pressure through a liquid to the working end of the tool?
   a. Hydraulic
   b. Pneumatic
   c. Hydroelectric
   d. Electro-pneumatic

_____ 11. Electrical circular saws should be rated _____.
   a. Light duty
   b. Medium duty
   c. Heavy duty
   d. Extra-heavy duty

_____ 12. Which type of blade is most likely to wear out quickly, chip, or become pinched from use on aircraft?
   a. Carbide-tipped
   b. Composite
   c. Serrated
   d. Diamond-tipped

_____ 13. A technique for quickly opening a large hole in the side of an aircraft is to use a(n) _____ in conjunction with a(n) _____.
   a. Air chisel; hydraulic cutter
   b. Hydraulic spreader; hydraulic cutter
   c. Come-along; hydraulic spreader
   d. Air chisel; electric winch

_____ 14. Pneumatic tools may be powered by all of the following *except* _____.
   a. Breathing apparatus cylinders
   b. Cascade systems
   c. Air systems on ARFF vehicles
   d. Compressed oxygen bottles

_____ 15. Which of the following tools increases the pulling ability of a lever to its maximum through a ratchet/pulley action?
   a. Come-along
   b. Truck-mounted winch
   c. Speed-wrench
   d. Hydraulic spreader

_____ 16. Which of the following tools is commonly used for personnel applications such as seats or slings?
   a. Come-along
   b. Rope
   c. Chain
   d. Webbing

_____ 17. Which of the following devices can be used to secure landing gear assemblies, canopy jettison systems, and seat ejection systems?
   a. Keys
   b. Pins
   c. Tumblers
   d. Bolts

_____ 18. One common way of designating collection points for equipment and personnel is to use salvage covers of different _____.
   a. Sizes
   b. Shape
   c. Fabrics
   d. Colors

7

_____ 19. Which of the following statements about air lifting bags is true?
   a. They are difficult to use in rescue and aircraft stabilization work.
   b. The working pressure often exceeds 500 psi *(3 500 kPa)*.
   c. They produce enough force to lift or displace enormous objects.
   d. They should not be stacked.

_____ 20. The easiest and quickest way for rescue personnel to gain access to an aircraft is through _____.
   a. Windows
   b. Forcible entry
   c. Doors and hatches
   d. Emergency openings

_____ 21. All of the following statements about aircraft windows are true *except* _____.
   a. Aircraft windows may be used for ventilation.
   b. Most aircraft windows that are modified for use as emergency exits have latch releases both outside and inside the cabin.
   c. Most window exits open toward the outside of the cabin.
   d. Personnel may use windows during rescue.

_____ 22. Which of the following statements about escape slides is true?
   a. Escape slides present only minor hazards to ARFF personnel.
   b. Escape slides inflate with enough force and pressure to cause severe injury or death to anyone in the immediate area.
   c. The slides are generally attached to the fuselage by rings and chains.
   d. All escape slides release automatically when the door opens.

_____ 23. Areas outlined with yellow or black dashed lines on military aircraft _____.
   a. Are designed for cutting
   b. Contain fuel
   c. Contain explosives or weapons
   d. Are automatic-opening escape hatches

_____ 24. If it is absolutely necessary to cut in order to enter an aircraft, where should the cutting be done?
   a. On the top of the fuselage
   b. Around or near windows
   c. On the underside of the fuselage
   d. At the doors

_____ 25. Reinforcements of an aircraft's skin are almost always _____.
   a. Unpredictable
   b. Diagonal to the length of the fuselage
   c. Irregular curves
   d. Parallel or perpendicular to the length of the fuselage

_____ 26. Which of the following statements about the area selected for cutting an aircraft is true?
   a. An area with a lot of rivets is preferable.
   b. The area should consist of one or more rectangular skin surfaces.
   c. The main deck is a good area for cutting.
   d. The area should be just below the main deck if possible.

Identify

C. Identify guidelines for lighting equipment. Write an *X* before each correct statement below.

_____ 1. ARFF personnel should know how to quickly set up portable lighting as well as how to operate all lighting sources on the apparatus.

_____ 2. Portable electric generators may be used to operate chain saws, circular saws, reciprocating saws, and other power tools.

_____ 3. Generators that are mounted on apparatus should be permanently fixed to prevent ground fault shocks.

_____ 4. Most airport fire departments do not need portable lighting units because the lights on apparatus will provide all the direct lighting needed in the ARFF environment.

_____ 5. Most modern crash apparatus have high-powered floodlights mounted on the front, on the sides, and on the rear.

_____ 6. Modern portable lighting equipment has built-in safeguards that prevent it from igniting a flammable atmosphere.

_____ 7. Personnel should treat all wires as if they are "hot" and of high voltage.

_____ 8. Personnel wearing full protective clothing may safely use tools that are not insulated.

_____ 9. It may be necessary to remove the third prong from a grounded electrical plug in order to attach it to an insulated tool.

_____ 10. Touching any tool that is in contact with live electrical wires will complete the circuit to ground, resulting in electrical shock.

Chapter 8: Aircraft Rescue and Fire Fighting Apparatus Driver/Operator

True/False

A. Write *True* or *False* before each of the following statements. Correct those statements that are false.

_____ 1. A minimum of two firefighters must be assigned to all ARFF apparatus.

_____ 2. Aircraft may have to be diverted from airports that fall below minimum vehicle requirements due to vehicle maintenance problems.

_____ 3. Handline delivery methods for ARFF apparatus may include 1½-inch or 1¾-inch *(38 mm or 45 mm)* hoselines, booster reels, and multiagent handlines.

_____ 4. The size of dry-chemical systems on ARFF apparatus is usually no greater than 500 pounds *(227 kg)*.

_____ 5. The only way to ensure that a dry-chemical system functions correctly is to activate it on some sort of time schedule.

_____ 6. The future production of all clean agents has been banned because they harm the earth's ozone layer.

7. Servicing clean-agent extinguishing systems is simple and safe enough for any untrained firefighter to perform.

8. Structural apparatus driver/operator training also qualifies personnel as ARFF apparatus driver/operators.

9. The tremendous weights of ARFF apparatus make them more stable than passenger vehicles or pickup trucks.

10. ARFF vehicles are designed to withstand over-acceleration and excessive deceleration without needing additional maintenance.

11. Automatic transmissions are far more efficient than manual transmissions.

12. ARFF vehicles are designed to function on steep grades and can be driven on grades the same as they are on flat ground.

13. Because ARFF vehicles are designed to allow for maximum ground clearance, they will *not* catch on objects in their paths.

14. ARFF apparatus driver/operators should keep windows closed when approaching a suspected accident/incident site in poor visibility.

_____ 15. Factors that determine aircraft wreckage patterns include direction and speed on impact, weather conditions, size of aircraft, type of crash, and location of the crash site.

_____ 16. Current agent management practice is to hurry to the aircraft, dump everything, and then rush back to resupply.

_____ 17. One way to learn an extinguishing system's reach is to practice knocking a softball off a traffic cone.

_____ 18. Each manufacturer has a different design for its ARFF vehicles' structural fire-fighting systems.

Multiple Choice

B. Write the letter of the best answer on the blank before each statement.

_____ 1. Who is responsible for safely transporting firefighters and apparatus to and from the scene of an ARFF emergency?
   a. Safety officer
   b. Chief mechanic
   c. Driver/operator
   d. Fire chief

_____ 2. Which NFPA standard sets minimum qualifications for ARFF apparatus driver/operators?
   a. NFPA 1001
   b. NFPA 1002
   c. NFPA 2001
   d. NFPA 2002

_____ 3. The main reason for vehicle inspection is to _____.
   a. Reduce maintenance costs.
   b. Meet warranty requirements.
   c. Ensure proper operation.
   d. Satisfy insurance provisions.

8

____ 4. A daily vehicle inspection should include setting the mirrors and safety equipment for ____.
   a. An average-sized individual
   b. The person performing the inspection
   c. The person operating the vehicle
   d. The most senior driver/operator on duty

____ 5. Many modern ARFF vehicles carry and dispense at least ____ types of extinguishing agents.
   a. Two
   b. Three
   c. Four
   d. Five

____ 6. Daily inspections of foam systems on ARFF apparatus usually consist of ____.
   a. Ensuring that the system is discharging the appropriate ratio of foam concentrate
   b. Checking the foam discharge pressure
   c. Inspecting all seals and gaskets and replacing as necessary
   d. Making sure that the agent tank is full

____ 7. What will happen if the foam-to-water mixture is too lean?
   a. Foam will be wasted.
   b. Foam bubbles will drain sooner than desired.
   c. Foam will lose viscosity.
   d. Foam will be too dry.

____ 8. According to NFPA standards, all of the following are acceptable methods of testing a foam proportioning system *except* ____.
   a. Foam concentrate reactivity
   b. Foam concentrate displacement
   c. Foam solution refractivity
   d. Foam solution conductivity

____ 9. The most common method of dispensing dry chemical from ARFF apparatus is to use ____.
   a. A piggyback system
   b. Water-stream injection
   c. Scoop shovels
   d. A handline

____ 10. Which dry-chemical dispensing method is especially effective for fighting three-dimensional fires because it also stops associated spill fires?
   a. Piggyback system
   b. Water-stream injection
   c. Scoop shovels
   d. Handline

____ 11. To prevent dry chemical from caking, older dry-chemical systems may require ____ on a regular basis.
   a. Replacing the chemical
   b. Rinsing the system
   c. Fluffing the chemical
   d. Adding absorbent materials

12. Substances that interrupt the chemical chain reaction of combustion without leaving any residue after they have been used are called _____ agents.
    a. Sterile
    b. Clean
    c. Immaculate
    d. Scrubbed

13. The first concern of safe vehicle operation before leaving the fire station is to _____.
    a. Secure all heavy objects in the cab.
    b. Adjust mirrors, seats, and steering column.
    c. Wear seat belts.
    d. Set the radio volume.

14. Braking reaction time can best be described as the time it takes _____.
    a. The driver/operator to react to a situation
    b. A vehicle to stop after the driver/operator applies the brakes
    c. The driver/operator to recognize a situation and bring a vehicle to a stop
    d. The brakes to begin taking effect after the driver/operator applies them

15. The last thing that most ARFF apparatus driver/operators remember when their vehicles roll over is _____.
    a. Applying the brakes
    b. Completing a turn
    c. Speeding up
    d. Making a turn

16. All new ARFF vehicles purchased with FAA funds must be equipped with a device that indicates to the driver/operator the vehicle's _____ when making a turn.
    a. Attitude
    b. Speed-to-braking distance ratio
    c. Critical turning radius
    d. Center of gravity

17. The faster a vehicle moves, the more _____ force is exerted when the vehicle is turned.
    a. Gravitational
    b. Centrifugal
    c. Radial
    d. Centripetal

18. Which of the following statements about skid avoidance is true?
    a. The great weight of ARFF vehicles makes them easier to stop.
    b. A good way to train for braking on frozen surfaces is to race toward an icy patch and attempt to control the vehicle.
    c. Understanding a vehicle's braking reactions on dry surfaces contributes little to the driver/operator's perception of how the vehicle will react on wet surfaces.
    d. The power needed by the vehicle drivetrain to move the vehicle at high speeds can make braking difficult.

8

_____ 19. Which of the following statements about driving ARFF vehicles on the airport is true?
   a. Only a small minority of the driving performed by driver/operators of ARFF vehicles is done on the airport.
   b. One of the most important driving situations is the route taken when responding to emergencies.
   c. Aircraft parking areas are relatively risk free for driver/operators of ARFF apparatus.
   d. The majority of responses to runway standby positions will change with each individual situation.

_____ 20. Which of the following statements about the Central Inflation/Deflation System is true?
   a. These systems are controlled from a control panel at the left rear wheel.
   b. Deflating tires makes them more likely to fill up with mud and dirt.
   c. These systems use an onboard air compressor.
   d. Inflating the tires increases traction on off-road surfaces.

_____ 21. The turnaround exercise required by NFPA _____ is excellent for giving the ARFF apparatus driver/operator experience in making these maneuvers.
   a. 1001
   b. 1002
   c. 2000
   d. 2001

_____ 22. As a rule of thumb, how should the ARFF apparatus driver/operator negotiate side slopes?
   a. Drive along a valley at right angles to its slope.
   b. Approach slopes head on.
   c. Do not drive on side slopes.
   d. Approach side slopes diagonally.

_____ 23. As pilots practice flying while using only instrumentation, the ARFF apparatus driver/operator should regularly practice driving with _____.
   a. The DEVS
   b. Headlights off
   c. Black-light goggles
   d. RADAR

_____ 24. To keep aircraft victims from approaching emergency vehicles, some departments have taken on the practice of _____.
   a. Shutting off the emergency lights
   b. Using flashing red warning lights
   c. Warning victims to stay clear over the loudspeaker system
   d. Using short blasts of the siren and horn

_____ 25. When positioning at a downed aircraft, a basic rule of thumb is to _____.
   a. Approach from the quickest route.
   b. Position uphill, upwind, and upstream of the incident site.
   c. Drive as close as possible.
   d. Position downhill, at crosswinds, and upstream of the incident site.

Aircraft Rescue and Fire Fighting

26. The most important aspect of agent management is _____.
    a. Understanding the agent's thermodynamic properties
    b. Keeping accurate records of the cost of agent used
    c. Knowing the amount of agent carried on the apparatus and its fire-fighting capabilities
    d. Accurate proportioning of the agent-water solution

27. Which of the following statements regarding wind and terrain is true?
    a. Airports tend to be less windy than more built-up locations.
    b. Wind is more of a problem with foam than with dry chemical or clean agents.
    c. The main concern regarding terrain is the flow of agent when applied on a flat, dry area.
    d. The problems of dry chemicals can be partially lessened by injecting the dry chemical into a water/foam stream.

28. A vehicle turret's capabilities are driven by all of the following *except* _____.
    a. Specific gallons (liters) per minute
    b. Storage tank capacities
    c. Pressure
    d. Turret location

29. The best way to make sure the agent is actually reaching the fire and extinguishing it is to _____.
    a. Discharge the agent in a steady flow.
    b. Discharge the agent in 60-second intervals.
    c. Discharge the agent for 15 seconds at a time with 5-second pauses.
    d. Discharge the agent a few seconds at a time.

30. All of the following statements about pump-and-roll are true *except* _____.
    a. Pumping and rolling exercises are too expensive, hazardous, and time-consuming to be practiced regularly.
    b. All major ARFF vehicles have the ability to pump and roll.
    c. Even with modern technology, turret control is difficult during pump-and-roll operations.
    d. Each day during morning checkout is a good time to practice pumping and rolling.

31. Which of the following statements about agent resupply is true?
    a. The most common agent resupply need during ARFF operations is the need to refill major ARFF vehicles with foam concentrate.
    b. ARFF apparatus' foam concentrate tanks are sized to supply approximately enough concentrate for one full tank of water.
    c. Dry-chemical or clean-agent systems are easier to resupply than foam systems.
    d. Each department should determine how much agent is needed to extinguish a fire in the largest aircraft expected to land at the airport.

32. Winterization systems on ARFF vehicles consist of _____.
    a. Insulation wraps
    b. Antifreeze injection systems
    c. Mini-heating systems
    d. Agitators and/or vibrators

33. Aircraft-skin-penetrating devices use _____ to penetrate the aircraft skin.
    a. Extreme-high-pressure water jets
    b. Penetrating nozzles
    c. Pneumatic hammers
    d. Laser beams

34. Which of the following statements about compressed-air foam systems is true?
    a. Their maximum air-to-agent ratio is 10:1.
    b. Their highly expanded foam has a shortened drain time.
    c. The higher air ratio makes handlines lighter and reduces extinguishment times.
    d. The higher air ratio reduces the quality of the foam blanket.

35. Onboard electrical generators on ARFF vehicles are generally used for _____.
    a. On-scene lighting
    b. Reserve pumping capacity
    c. Communications equipment
    d. Powering aircraft systems

Chapter 9 Extinguishing Agents

Matching

A. Match to their definitions terms associated with extinguishing agents. Write the appropriate letters on the blanks.

____ 1. Raw foam liquid before the introduction of water and air

____ 2. Device that introduces foam concentrate into the water stream

____ 3. Mixture of foam concentrate and water before the introduction of air

____ 4. Completed product after air is introduced into the foam solution

____ 5. Petroleum-based fuel

____ 6. Flammable liquid that can be mixed in water

____ 7. Creating a barrier between the fuel and the fire

____ 8. Preventing the release of flammable vapors

____ 9. Mixing of water with foam concentrate to form a foam solution

____ 10. Uses the pressure energy in the stream of water to draft foam concentrate into the fire stream

____ 11. Uses an external pump or head pressure to force foam concentrate into the fire stream

____ 12. Increase in volume of a foam solution when it is aerated

____ 13. Mixed with air

a. Expansion
b. Foam solution
c. Polar solvent
d. Induction
e. Vaporizing
f. Aerated
g. Suppressing
h. Foam
i. Foam concentrate
j. Hydrocarbon
k. Injection
l. Separating
m. Foam proportioner
n. Proportioning

75
Aircraft Rescue and Fire Fighting

9

True/False

B. Write *True* or *False* before each of the following statements. Correct those statements that are false.

_____ 1. ARFF personnel could encounter all four classes of fire in any one incident.

_____ 2. The U.S. Department of Transportation considers all aircraft fuels to be Class 3 hazards.

_____ 3. Blended fuels such as Jet-B have a higher ignition temperature than Jet-A fuels.

_____ 4. When using water to extinguish aircraft fires, firefighters have been most successful when they have used straight streams.

_____ 5. Because of the loss of water through evaporation, foam may have to be reapplied to a burning surface to be effective.

_____ 6. Technological advances have made the use of foam foolproof.

_____ 7. Class B foams designed solely for hydrocarbon fires will also extinguish polar solvent fires if they are applied at high enough concentrations.

_____ 8. Foam proportioners may be used with any delivery device made by the same manufacturer.

_____ 9. Premixed foam proportioning systems are limited to a one-time application.

_____ 10. Most ARFF apparatus are equipped with integral, onboard foam-proportioning systems.

_____ 11. Class B foam concentrates of similar type but from different manufacturers may be mixed together if they are manufactured to U.S. military specifications.

_____ 12. Foam should be applied to unignited spills and ignited spills at the same rate.

_____ 13. AFFF can "heal" areas where the foam blanket is disturbed.

_____ 14. Regular protein foam is widely used in aircraft fire fighting because it is noncorrosive and self-sealing.

_____ 15. The in-line eductor is the most common type of foam proportioner used in ARFF applications.

_____ 16. Use of a foam nozzle eductor compromises firefighters' safety because they cannot move quickly and must leave their concentrates behind if they are required to withdraw for any reason.

_____ 17. Apparatus-mounted foam proportioning systems are rarely found on ARFF apparatus.

_____ 18. A major disadvantage of old around-the-pump proportioners is that the pump cannot take advantage of incoming pressure.

_____ 19. High-energy foam generating systems generally differ from other systems in that they introduce compressed air into the foam solution *prior* to discharge into the hoseline.

_____ 20. Most ARFF apparatus use high-energy foam generating systems.

_____ 21. Standard fire-fighting nozzles can be used for applying some types of low-expansion foams.

_____ 22. Many jurisdictions make a fire attack with standard fog nozzles and then switch to air-aspirating foam nozzles to blanket the fuel once the fire is extinguished.

_____ 23. The initial foam application at an aircraft fire should be to isolate the fuselage from the fire.

_____ 24. The terms *dry chemical* and *dry powder* are interchangeable.

_____ 25. All dry-chemical agents are nonconductive, making them suitable for use on energized electrical equipment.

_____ 26. Dry-chemical agents are nontoxic and are generally considered quite safe to use.

_____ 27. Multipurpose A:B:C-rated dry-chemical extinguishers are recommended for extinguishing aircraft engine fires.

_____ 28. Wheeled dry-chemical extinguishers store the extinguishing agent and the pressurizing gas in the same tank.

_____ 29. The top of a wheeled dry-chemical extinguisher should be pointed away from the firefighter or other personnel when pressurizing the unit.

_____ 30. The *Montreal Protocol on Substances that Deplete the Ozone Layer* required a complete phaseout of the production of halogens by the year 2000.

9

Multiple Choice

C. Write the letter of the best answer on the blank before each statement.

____ 1. Which of the following is *not* one of the three basic types of fuels used in aircraft?
   a. AVGAS
   b. Formula I
   c. Kerosene
   d. Blends of gasoline and kerosene

____ 2. AVGAS has an octane rating of ____.
   a. 87–92
   b. 92–115
   c. 100–145
   d. 125–150

____ 3. Which of the following statements about aircraft fuels is true?
   a. Spills of low flash point fuels over 10 feet *(3 m)* in any direction and covering an area of over 50 square feet *(4.6 m²)* should be blanketed with foam.
   b. Blends are the most common grades of jet fuel.
   c. In general, jet fuels have lower flash points than AVGAS.
   d. Firefighters at fuel spills are safe as long as the ambient air temperature remains well below the flash point of the fuel.

____ 4. The autoignition temperature of Jet A-1 is ____.
   a. 475–500°F *(246°C to 260°C)*
   b. 440–475°F *(227°C to 246°C)*
   c. 442°F *(228°C)*
   d. 410–475°F *(210°C to 246°C)*

____ 5. If a fire has not occurred soon after an aircraft crashes, how should firefighters approach the aircraft?
   a. As though a fire is unlikely
   b. As though a fire is moderately likely
   c. As though a fire is highly likely
   d. As though it is on fire

____ 6. To prevent flashback at an aircraft accident, firefighters should ____.
   a. Shut down all aircraft electrical systems.
   b. Completely cover fuel-saturated areas with foam.
   c. Channel all spilled fuel to underwater storage tanks.
   d. Cover spilled fuel with absorbent materials.

____ 7. Which types of extinguishing agents are designed for mass application and rapid knockdown?
   a. Primary
   b. Secondary
   c. Class A
   d. Auxiliary

_____ 8. Which of the following statements about extinguishing agents is true?
   a. Foaming agents are the primary agents used to combat three-dimensional fires.
   b. Most fires involving aircraft will require auxiliary agents.
   c. For an auxiliary agent to be compatible with a primary agent, the auxiliary agent must rapidly break down foam.
   d. Generally, auxiliary agents are *not* effective as primary agents because they are prone to flashback.

_____ 9. Fires involving combustible metals must be extinguished using a(an) _____ combustible-metal extinguishing agent.
   a. Dry-powder
   b. Halon
   c. AFFF
   d. Water-soluble

_____ 10. The preferred extinguishing agent for fires in the interior of aircraft involving Class A materials is _____.
   a. Water
   b. Foam
   c. Dry chemical
   d. Halon

_____ 11. Water can be used for all of the following *except* to _____.
   a. Push burning fuel away from the aircraft.
   b. Cool the aircraft fuselage.
   c. Provide a heat shield for aircraft passengers who are evacuating and personnel who are fighting the fire.
   d. Extinguish burning hydrocarbon fuels.

_____ 12. Firefighters should avoid using _____ water streams because they tend to churn and splash the fuel.
   a. Straight
   b. Broken
   c. Fog
   d. Spray

_____ 13. Which type of extinguishing agent has a lower specific gravity than aircraft fuels?
   a. Water
   b. Foam
   c. Dry chemical
   d. Halon

_____ 14. The specifically defined area within which it is feasible to extinguish or control an aircraft fire long enough for ARFF personnel to rescue trapped or immobilized passengers is the _____.
   a. Safety zone
   b. Fire-control area
   c. Damage-control perimeter
   d. Maximum-attack sector

_____ 15. Which of the following is a hydrocarbon fuel?
   a. Alcohol
   b. Naphtha
   c. Ketones
   d. Lacquer thinner

___ 16. Which of the following is a polar solvent fuel?
   a. Benzene
   b. Jet fuel
   c. Esters
   d. Gasoline

___ 17. In general, foam works by _____.
   a. Forming a blanket on the burning fuel
   b. Penetrating the burning fuel
   c. Breaking up the burning fuel
   d. Altering the chemical structure of the burning fuel

___ 18. Most fire-fighting foam concentrates are intended to be mixed with _____ water.
   a. 94 to 99.9 percent
   b. 90 to 95 percent
   c. 88.8 to 99.9 percent
   d. 87 to 94.9 percent

___ 19. A disadvantage of batch mixing is that it _____.
   a. Is complicated
   b. Does *not* allow accurate proportioning
   c. Requires shutting down foam attack lines when refilling the tank
   d. Can be used only for mixing foam in portable tanks

___ 20. All of the following are common methods of foam concentrate storage *except* _____.
   a. Pails
   b. Stationary storage tanks
   c. Barrels
   d. Apparatus tanks

___ 21. Which of the following statements regarding Class A foam is true?
   a. Class A foam is a new technology.
   b. Class A foam has yet to come of age.
   c. Class A foams are also effective in fighting Class B fires.
   d. Class A foams have limited applications in ARFF settings.

___ 22. Which of the following types of foam may be applied with standard fog nozzles?
   a. Aqueous film forming foam
   b. Regular protein foam
   c. Fluoroprotein foam
   d. High-expansion foam

___ 23. Multipurpose Class B foams are normally used at a _____ rate on hydrocarbons and at a _____ rate on polar solvents.
   a. 1 percent; 3 percent
   b. 3 percent; 6 percent
   c. 1 percent or 3 percent; 3 percent or 6 percent
   d. 3 percent or 6 percent; 6 percent or 10 percent

24. Medium-expansion foam has an air/solution ratio of ____.
    a. Up to 20:1
    b. 20:1 to 100:1
    c. 20:1 to 200:1
    d. 100:1 to 200:1

25. The minimum foam solution application rates for aircraft fuel spill fires are established in ____.
    a. NFPA 11
    b. NFPA 403
    c. NFPA 405
    d. NFPA 414

26. Which of the following types of foam is the recommended extinguishing agent for hydrocarbon fuel fires?
    a. FFFP
    b. PF
    c. High-expansion
    d. AFFF

27. Which of the following statements about AFFF is true?
    a. On most polar solvents, alcohol-resistant AFFF is used at a 1-percent or 3-percent proportion.
    b. AFFF must be used only with freshwater.
    c. Alcohol-resistant AFFF *cannot* be used on hydrocarbon fires.
    d. AFFF may be applied with either an aspirating or nonaspirating nozzle.

28. Which type of foam is widely used in protecting fuel tanks because of its unique fuel-shedding qualities?
    a. Aqueous film forming foam
    b. Regular protein foam
    c. Fluoroprotein foam
    d. High-expansion foam

29. Which of the following statements about film forming fluoroprotein foam is true?
    a. It is based on FPF technology with AFFF capabilities.
    b. It is ineffective on flammable liquid fires.
    c. It is suitable for use only with freshwater.
    d. It is more effective than AFFF in maintaining foam stability.

30. All of the following are basic applications of high-expansion foam *except* ____.
    a. Concealed spaces
    b. Class A fire applications
    c. Class B fire applications
    d. Fixed extinguishing systems

31. A low-energy foam proportioning system imparts pressure on the foam solution solely by the use of a ____.
    a. Gas turbine
    b. Fire pump
    c. Air compressor
    d. Hand pump

9

_____ 32. For the nozzle and in-line foam eductor to operate properly, both must have the same _____.
   a. Part number
   b. Letter rating
   c. Manufacturer
   d. Rating in gpm *(L/min)*

_____ 33. The self-educting master stream foam nozzle is used where flows in excess of _____ are required.
   a. 250 gpm *(1 000 L/min)*
   b. 350 gpm *(1 400 L/min)*
   c. 450 gpm *(1 700 L/min)*
   d. 550 gpm *(2 100 L/min)*

_____ 34. The only difference between installed in-line eductors and portable in-line eductors is that _____.
   a. Portable in-line eductors have higher capacities.
   b. Installed in-line eductors require fewer precautions.
   c. Installed in-line eductors operate at higher pressures.
   d. Installed in-line eductors are permanently attached to the apparatus pumping system.

_____ 35. Proportioning systems that consist of an in-line eductor on a small return water line connected from the discharge side of the pump back to the intake side of the pump are called _____ proportioners.
   a. Free-cycle
   b. Pump-detour
   c. Around-the-pump
   d. Infinite-circle

_____ 36. Which type of apparatus-mounted proportioning system is rated for a specific flow and should be used at that rate?
   a. Installed in-line eductor
   b. Balanced-pressure
   c. Emulsifier
   d. Around-the-pump

_____ 37. Which types of foam proportioning systems provide the most accurate proportioning of foam concentrate at large flow rates?
   a. Installed in-line eductor
   b. Balanced-pressure
   c. Emulsifier
   d. Around-the-pump

_____ 38. An advantage of bypass-type balanced-pressure proportioners is that _____.
   a. They can monitor the demand for foam concentrate and adjust the amount supplied.
   b. The bypass of concentrate causes foam concentrate aeration.
   c. They do *not* require a separate foam pump with PTO or other power source.
   d. There is *no* recirculation back to the foam concentrate tank.

_____ 39. Which type of proportioning system uses a foam concentrate pump that supplies foam concentrate to a venturi-type proportioning device built into the water line?
   a. Bypass-type balanced-pressure
   b. Direct-stream balanced-pressure
   c. Variable-flow demand-type balanced-pressure
   d. Direct injection

_____ 40. A disadvantage of variable-flow demand-type balanced-pressure proportioning systems is that _____.
   a. Pressure drops across the discharge are higher than those on standard pumpers.
   b. Foam concentrate recirculates back into the foam concentrate tank.
   c. The system must be flushed after each use.
   d. Foam concentrate flow and pressure do *not* match system demand.

_____ 41. Which type of proportioning system controls the foam concentrate ratio by monitoring the water flow and controlling the speed of a positive-displacement foam concentrate pump?
   a. Bypass-type balanced-pressure
   b. Direct-stream balanced-pressure
   c. Variable-flow demand-type balanced-pressure
   d. Variable-flow variable-rate direct injection

_____ 42. The disadvantage of variable-flow variable-rate direct-injection systems is that _____.
   a. They do *not* automatically adjust to changes in water flow when nozzles are closed.
   b. Nozzles must remain below the pump or proportioning will be affected.
   c. They *cannot* be used with high-energy foam systems.
   d. They *cannot* discharge foam and water from different outlets at the same time.

_____ 43. Which of the following is an advantage of the batch-mixing method of proportioning foam?
   a. It is simple.
   b. It allows for continuous discharge on large incidents.
   c. Maintaining the correct concentrate ratio is easy, even if the tank is *not* completely empty.
   d. It allows foam and water to be discharged from different outlets at the same time.

_____ 44. According to IFSTA's definition, a handline nozzle flows less than _____.
   a. 250 gpm *(1 000 L/min)*   c. 450 gpm *(1 700 L/min)*
   b. 350 gpm *(1 400 L/min)*   d. 550 gpm *(2 100 L/min)*

9

_____ 45. The best foam application for fog nozzles is with _____.
   a. FPF
   b. AFFF
   c. FFFP
   d. High-expansion foam

_____ 46. Which of the following statements about handline nozzles is true?
   a. Fog nozzles provide maximum expansion of the agent.
   b. Air-aspirating foam nozzles have considerably greater reach than fog nozzles.
   c. Fog nozzles are best suited for attacks on Class C fires.
   d. Air-aspirating foam nozzles are best suited for applying a foam blanket to unignited or recently extinguished pools of fuel.

_____ 47. Which of the following statements about turret nozzles is true?
   a. All turret nozzles are aspirating.
   b. Nonaspirating turrets produce better quality foam than aspirating turrets.
   c. The type of turret selected is simply a question of preference and local need.
   d. Nonaspirating turrets have a shorter reach than aspirating turrets.

_____ 48. All of the following are common reasons for failure to generate foam or for generating poor-quality foam *except* for _____.
   a. Failure to match eductor and nozzle flow
   b. Improper cleaning of proportioning equipment
   c. Hose lay being too long on the intake side of the eductor
   d. Nozzle being too far above eductor

_____ 49. Which type of foam can be used in nonaspirating applications?
   a. FPF
   b. AFFF
   c. FFFP
   d. High-expansion foam

_____ 50. Nonaspirating devices generally achieve an expansion ratio of _____.
   a. 1:1 or 2:1
   b. 2:1 or 3:1
   c. 3:1 or 6:1
   d. 6:1 or 10:1

_____ 51. Aspirating devices generally achieve an expansion ratio of _____.
   a. 1:1 or 2:1
   b. 2:1 or 3:1
   c. 3:1 or 6:1
   d. 6:1 or 10:1

_____ 52. The general tactics of ARFF foam application are to _____.
   a. Insulate and isolate
   b. Protect and preserve
   c. Search and destroy
   d. Contain and control

_____ 53. The ideal angle for viewing the effectiveness of a turret's reach is _____.
   a. 45 degrees
   b. 60 degrees
   c. 90 degrees
   d. 180 degrees

54. Which foam application method directs the foam stream on the ground near the front edge of a burning liquid pool?
    a. Seat-of-the-fire
    b. Bank-down
    c. Rain-drop
    d. Roll-on

55. Which foam application method is used when an elevated object is near or within the area of a burning liquid pool or unignited liquid spill?
    a. Seat-of-the-fire
    b. Bank-down
    c. Rain-drop
    d. Roll-on

56. Which foam application angle provides maximum effectiveness?
    a. 0 degrees
    b. 15 degrees
    c. 30 degrees
    d. 45 degrees

57. Dry-chemical agents are effective for all of the following applications *except* _____.
    a. Initially attacking hydraulic fires
    b. Preventing flashback
    c. Attacking three-dimensional fires
    d. Extinguishing running-fuel fires

58. Which of the following is a correct guideline for applying dry chemicals?
    a. Apply dry chemicals from a position at crosswinds to the fire when possible.
    b. Avoid blanketing the fire with dry-chemical agents.
    c. Aggressively splash and churn the fuel.
    d. Use dry chemicals in conjunction with film forming foam.

59. Stored-pressure-type handheld dry-chemical extinguishers contain a constant pressure of about _____.
    a. 100 psi *(700 kPa)*
    b. 150 psi *(1 050 kPa)*
    c. 200 psi *(1 400 kPa)*
    d. 250 psi *(1 725 kPa)*

60. What type of cartridge do cartridge-operated dry-chemical extinguishers use if they are going to be exposed to freezing temperatures?
    a. Carbon dioxide
    b. Hydrogen
    c. Nitrous oxide
    d. Nitrogen

61. Which two halogenated agents are most commonly used for extinguishing fires?
    a. Halon 1012 and Halon 1311
    b. Halon 1200 and Halon 3001
    c. Halon 1211 and Halon 3100
    d. Halon 1211 and Halon 1301

62. Halons extinguish fires best as _____.
    a. Liquids
    b. Gases
    c. Mixtures of liquid and gas
    d. Vapor clouds

9

_____ 63. Which of the following statements about halogenated agents is true?
   a. Halogenated agents are ineffective in extinguishing Class B fires.
   b. Halons easily penetrate inaccessible areas.
   c. Halons leave a mildly corrosive residue.
   d. A halogenated agent's primary modern-day application is the protection of internal combustion engines.

Identify

D. Identify the following abbreviations associated with extinguishing agents. Write the correct interpretation before each.

_____ 1. AFFF

_____ 2. PF

_____ 3. FPF

_____ 4. FFFP

_____ 5. CAFS

Chapter 10 Aircraft Rescue and Fire Fighting Tactical Operations

True/False

A. Write *True* or *False* before each of the following statements. Correct those statements that are false.

_____ 1. The initial unit on the scene should transmit a clear report of conditions, summon whatever additional resources may be needed, and describe the plan of action to be implemented.

_____
_____

_____ 2. Because ARFF apparatus often respond single file, the first fire apparatus to the accident site may dictate the ultimate fire-fighting positions of other apparatus.

_____
_____

_____ 3. Gullies and downslope depressions near a crashed aircraft make good, protected positions for apparatus.

_____
_____

_____ 4. If evacuation has begun from the interior of the aircraft, ARFF personnel should turn their attention to extinguishment.

_____
_____

_____ 5. On propeller-driven aircraft, the prop wash may be used to help cool the wheels.

_____
_____

_____ 6. When a good tire is heated, the increase in air pressure alone is enough to cause it to fail.

_____
_____

_____ 7. Wheel fires may *not* occur until the aircraft is standing still because of the lag time for wheel parts to absorb enough brake heat for ignition.

_____ 8. Each wing tank of a Boeing® 757 holds 14,600 pounds *(6 623 kg)* of fuel, or just over 2,100 gallons *(8 400 L)*.

_____ 9. Small or medium static fuel spills contain such a small amount of fuel that they may be absorbed, picked up, and placed easily in an approved container.

_____ 10. The fire department should be summoned immediately to any fuel spill over 10 feet *(3 m)* in any dimension or over 50 square feet *(4.6 m^2)* in area.

_____ 11. The severity of the hazard created by a fuel spill depends primarily upon the amount of fuel involved.

_____ 12. Any cargo, baggage, mailbags, or similar items that have come in contact with fuel should be incinerated.

_____ 13. The usual method of controlling a tail cone fire is to turn off the fuel and motor (rev) the engine.

_____ 14. In aircraft, any indication of a fire or overheat condition in a wheel well after takeoff is cause for an immediate return and emergency landing.

_____ 15. Backdraft is a likely hazard in aircraft fires.

_____ 16. Firefighters should *not* attempt to ventilate an aircraft until all occupants have escaped.

_____ 17. Radioactive materials must be accessible by the flight crew and are often stored near the front of the aircraft.

_____ 18. Whenever a fire warning light activates while in flight, the crew immediately notifies air traffic control and ARFF personnel initiate an emergency response.

_____ 19. Even in low-impact crashes, ARFF personnel should initiate extrication operations only after donning full protective clothing and SCBA.

_____ 20. Crashes resulting from rejected takeoff with runway overrun are usually survivable.

10

_____ 21. Aircraft crash fires on hillsides sometimes may limit fire-fighting efforts to preventing fire spread and performing a thorough overhaul.

_____ 22. Promptness is the primary consideration in aircraft emergency responses.

_____ 23. Preserving the accident scene and safeguarding evidence is the responsibility of all ARFF personnel.

_____ 24. Emergencies that occur without prior warning are called unanticipated accidents.

_____ 25. The IC at an aircraft emergency must maintain a clear distinction between rescue and extinguishment activities.

_____ 26. Given an adequate supply of AFFF and available additional water, structural apparatus can sustain an effective aircraft fire attack as long as necessary.

_____ 27. In aircraft accidents involving fire, the initial attack generally involves operating both roof and bumper turrets while additional units stand by.

_____ 28. Cutting ventilation openings in an aircraft fuselage is highly recommended because it can reduce fire spread and can be done safely from an elevated platform.

_____ 29. Rescue apparatus should be positioned as close to the triage area as possible.

_____ 30. On civilian airliners, the lap belt release mechanism also releases a crew member's shoulder harness strap.

_____ 31. If an injured pilot is wearing a parachute and has a suspected back injury, the parachute should be left on to support the pilot's back whenever possible.

_____ 32. A thorough overhaul inspection at aircraft incidents is necessary only if fire was apparent.

_____ 33. A military aircraft crash site contains hazards identical to those of a civilian crash site.

_____ 34. Many military aircraft use a varied mixture of jet fuel, which has a significantly higher flash point than civil aviation fuel.

10

_____ 35. An area should be established where the walking wounded could meet with members of the press and attorneys.

_____

_____

Multiple Choice

B. **Write the letter of the best answer on the blank before each statement.**

_____ 1. Fire may burn through an aircraft's skin in as little as _____.
   a. 45 seconds
   b. 60 seconds
   c. 75 seconds
   d. 90 seconds

_____ 2. One of the most difficult decisions following the size-up may be whether to _____.
   a. Conserve agent during suppression operations.
   b. Take *no* action other than beginning to establish a command structure.
   c. Call for additional resources.
   d. Assume the position of Incident Commander.

_____ 3. If the first vehicle on the scene has taken a position that affords the route of greatest safety for the aircraft occupants, how should later arriving units approach the scene?
   a. Emergency lights and audible devices on
   b. Emergency lights on and audible devices off
   c. Emergency lights off and audible devices on
   d. Emergency lights and audible devices off

_____ 4. All of the following statements about wind direction are true *except* _____.
   a. When operating against the wind, the smoke obscures vision and the heat is more intense.
   b. Upwind paths of egress are safer for the aircraft's occupants because the exit corridor is free of heat and smoke.
   c. Attacking a fire from downwind should be attempted if it offers an approach that is significantly easier than any other.
   d. Attacking with the wind enables extinguishing agents to be applied more effectively, thus reducing extinguishing time.

_____ 5. Initial fire-fighting efforts should be directed at _____.
   a. A portion of the fuselage
   b. A section of a wing
   c. The fuel tanks
   d. The engine(s)

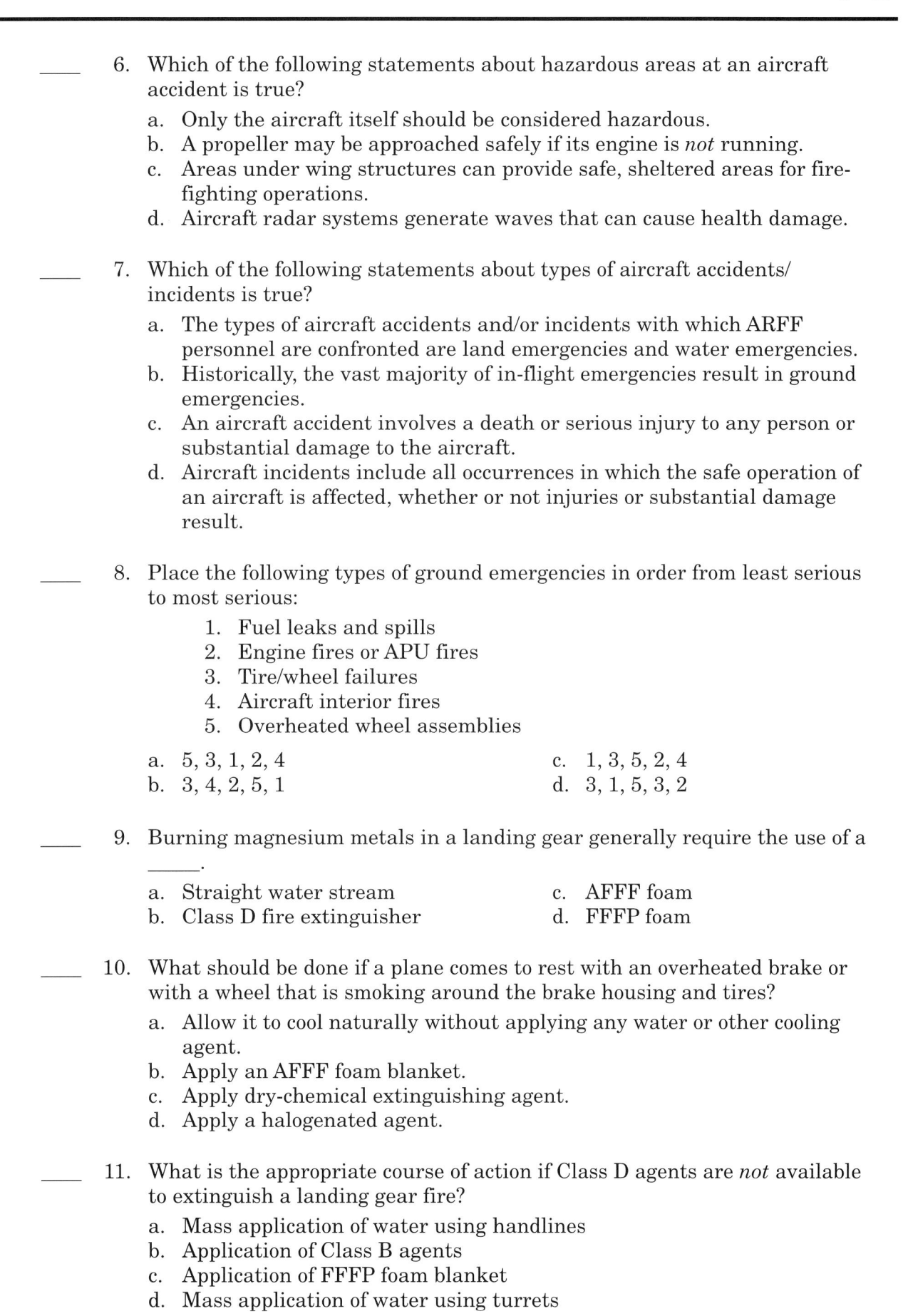

_____ 6. Which of the following statements about hazardous areas at an aircraft accident is true?
   a. Only the aircraft itself should be considered hazardous.
   b. A propeller may be approached safely if its engine is *not* running.
   c. Areas under wing structures can provide safe, sheltered areas for firefighting operations.
   d. Aircraft radar systems generate waves that can cause health damage.

_____ 7. Which of the following statements about types of aircraft accidents/incidents is true?
   a. The types of aircraft accidents and/or incidents with which ARFF personnel are confronted are land emergencies and water emergencies.
   b. Historically, the vast majority of in-flight emergencies result in ground emergencies.
   c. An aircraft accident involves a death or serious injury to any person or substantial damage to the aircraft.
   d. Aircraft incidents include all occurrences in which the safe operation of an aircraft is affected, whether or not injuries or substantial damage result.

_____ 8. Place the following types of ground emergencies in order from least serious to most serious:
   1. Fuel leaks and spills
   2. Engine fires or APU fires
   3. Tire/wheel failures
   4. Aircraft interior fires
   5. Overheated wheel assemblies

   a. 5, 3, 1, 2, 4
   b. 3, 4, 2, 5, 1
   c. 1, 3, 5, 2, 4
   d. 3, 1, 5, 3, 2

_____ 9. Burning magnesium metals in a landing gear generally require the use of a _____.
   a. Straight water stream
   b. Class D fire extinguisher
   c. AFFF foam
   d. FFFP foam

_____ 10. What should be done if a plane comes to rest with an overheated brake or with a wheel that is smoking around the brake housing and tires?
   a. Allow it to cool naturally without applying any water or other cooling agent.
   b. Apply an AFFF foam blanket.
   c. Apply dry-chemical extinguishing agent.
   d. Apply a halogenated agent.

_____ 11. What is the appropriate course of action if Class D agents are *not* available to extinguish a landing gear fire?
   a. Mass application of water using handlines
   b. Application of Class B agents
   c. Application of FFFP foam blanket
   d. Mass application of water using turrets

10

____ 12. Fusible plugs in modern aircraft wheels help reduce the possibility of wheel collapse and fragmentation by _____.
   a. Dousing the tire with water
   b. Releasing tire pressure
   c. Releasing a dry-chemical extinguishing agent
   d. Increasing tire pressure

____ 13. The line of probable fragmentation when a wheel shatters and explodes is at least _____.
   a. 100 feet *(30 m)* directly to each side of the heated wheel
   b. 250 feet *(76 m)* in all directions from the heated wheel
   c. 300 feet *(100 m)* directly to each side of the heated wheel
   d. 400 feet *(122 m)* to the front and back of the heated wheel

____ 14. Wheel assemblies of current generation jet airliners are primarily constructed of _____ alloy.
   a. Magnesium   c. Aluminum
   b. Titanium   d. Steel

____ 15. What should firefighters do if a tire reignites after being extinguished with dry-chemical agents?
   a. Apply additional dry-chemical each time flame reappears.
   b. Apply AFFF foam.
   c. Withdraw and attack the flame with a turret nozzle.
   d. Apply CO_2.

____ 6. Dry-chemical fire extinguishers are recommended for controlling tire fires because they are _____.
   a. Less likely to cause corrosive damage
   b. Less subject to the wind
   c. Less expensive to operate
   d. Less likely to create localized cooling

____ 17. If fusible plugs have deflated all the tires, water may be applied in _____.
   a. A steady stream   c. Short, intermittent bursts
   b. Long, evenly spaced intervals   d. A continuous fog pattern

____ 18. Which of the following metals is used on most large propeller-driven aircraft and in early transport-type jet aircraft landing gear, engine mountings, and wheel cover plates?
   a. Titanium   c. Carborundum
   b. Magnesium   d. Stainless steel

____ 19. Titanium is about _____ as heavy as steel.
   a. 56 percent   c. 100 percent
   b. 72 percent   d. 117 percent

____ 20. Which of the following is a specialized extinguishing agent for controlling magnesium and titanium fires?
   a. Titanic-X
   b. Herculaneum
   c. G-1 powder
   d. FFFP foam

____ 21. If specialized extinguishing agents are *not* available, what is the next best method of controlling magnesium or titanium fires?
   a. Water in straight streams
   b. Water in heavy, coarse streams
   c. Water in a fog pattern
   d. A blanket of AFFF foam

____ 22. If a fuel leak develops during aircraft servicing, the first step should be to _____.
   a. Notify safety personnel.
   b. Evacuate nonessential personnel from the area.
   c. Stop the fueling operation.
   d. Determine the cause of the leak.

____ 23. Which of the following statements about incidents involving fuel leaks or spills is true?
   a. Occupants should *not* be allowed to leave the aircraft until all possible ignition sources have been eliminated.
   b. Maintenance personnel must check for flammable vapors that may have entered concealed compartments.
   c. Maintenance records should describe the cause and assign blame for each incident.
   d. All electrical power to the aircraft except that required for radio communications and onboard extinguishing systems should be shut down.

____ 24. Small spills involving an area less than _____ in any plane dimension normally involve minor danger.
   a. 18 inches *(450 mm)*
   b. 36 inches *(915 mm)*
   c. 72 inches *(1 830 mm)*
   d. 144 inches *(3 660 mm)*

____ 25. Which of the following actions should be taken at small or medium static fuel spills?
   a. Cover the spill with a heavy foam blanket.
   b. Disperse the fuel into storm drains.
   c. Cover the spill with an approved dry-chemical extinguishing agent.
   d. Post a fire watch.

____ 26. What action should personnel take if spilled fuel enters sanitary sewers or storm drains?
   a. Dam inlets to prevent additional fuel from entering.
   b. Flush the sewers or drains with large amounts of water.
   c. Disperse the fuel back out of the sewer or drain if possible.
   d. Dispense dry-chemical extinguishing agents into the sewer or drain.

10

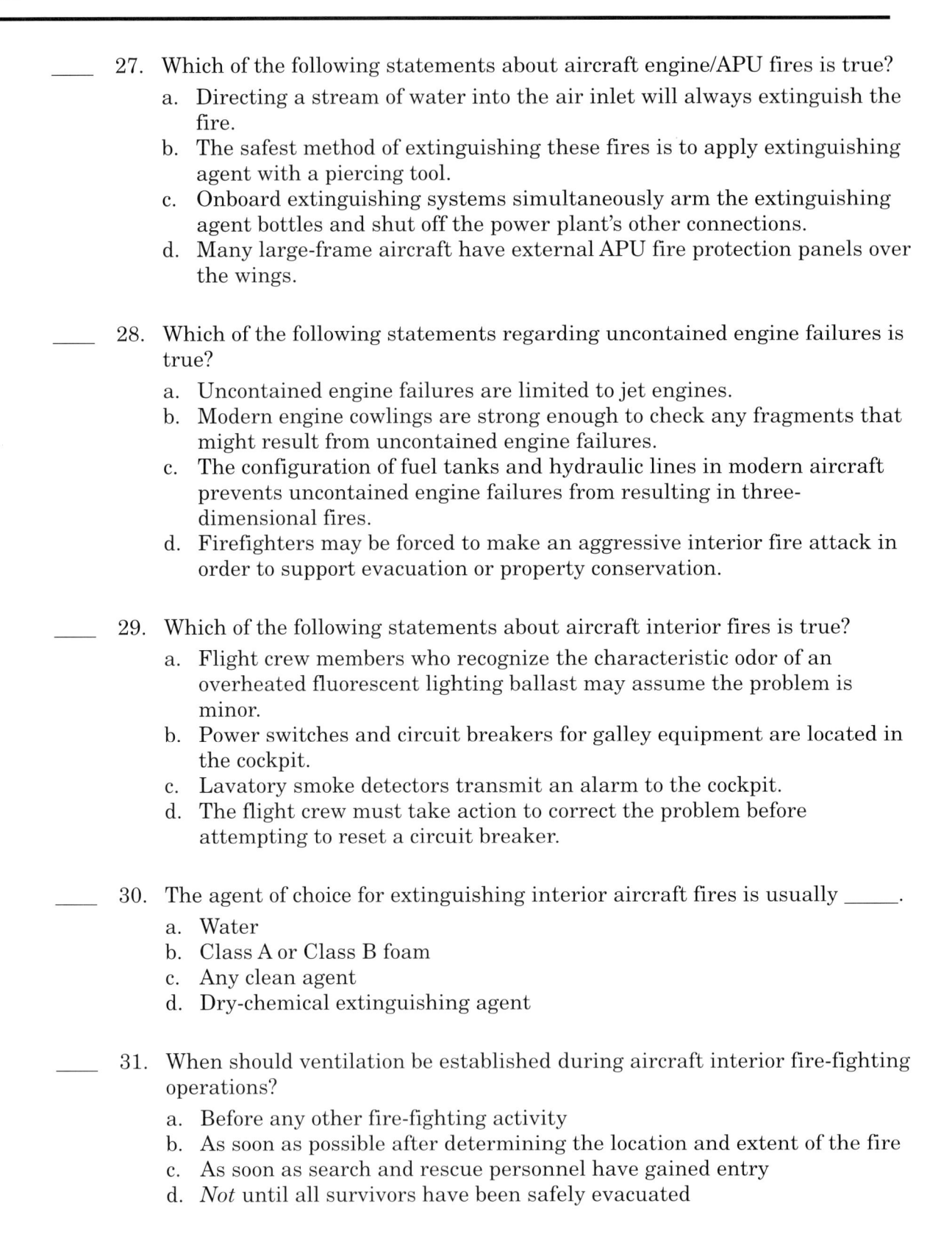

____ 27. Which of the following statements about aircraft engine/APU fires is true?
   a. Directing a stream of water into the air inlet will always extinguish the fire.
   b. The safest method of extinguishing these fires is to apply extinguishing agent with a piercing tool.
   c. Onboard extinguishing systems simultaneously arm the extinguishing agent bottles and shut off the power plant's other connections.
   d. Many large-frame aircraft have external APU fire protection panels over the wings.

____ 28. Which of the following statements regarding uncontained engine failures is true?
   a. Uncontained engine failures are limited to jet engines.
   b. Modern engine cowlings are strong enough to check any fragments that might result from uncontained engine failures.
   c. The configuration of fuel tanks and hydraulic lines in modern aircraft prevents uncontained engine failures from resulting in three-dimensional fires.
   d. Firefighters may be forced to make an aggressive interior fire attack in order to support evacuation or property conservation.

____ 29. Which of the following statements about aircraft interior fires is true?
   a. Flight crew members who recognize the characteristic odor of an overheated fluorescent lighting ballast may assume the problem is minor.
   b. Power switches and circuit breakers for galley equipment are located in the cockpit.
   c. Lavatory smoke detectors transmit an alarm to the cockpit.
   d. The flight crew must take action to correct the problem before attempting to reset a circuit breaker.

____ 30. The agent of choice for extinguishing interior aircraft fires is usually ____.
   a. Water
   b. Class A or Class B foam
   c. Any clean agent
   d. Dry-chemical extinguishing agent

____ 31. When should ventilation be established during aircraft interior fire-fighting operations?
   a. Before any other fire-fighting activity
   b. As soon as possible after determining the location and extent of the fire
   c. As soon as search and rescue personnel have gained entry
   d. *Not* until all survivors have been safely evacuated

____ 32. Which of the following statements regarding fires in unoccupied aircraft is true?
   a. These fires rarely develop into major incidents.
   b. As in structural fire fighting, opening an aircraft door under heavy smoke conditions requires vertical ventilation.
   c. Penetrating nozzles usually are ineffective in these situations.
   d. FAA requirements prohibit leaving aircraft attached to jetways during overnight layovers.

____ 33. The use of _____ may be the best tactic to combat a cargo aircraft interior fire.
   a. Skin-penetrating nozzles
   b. Extendable turrets
   c. Forcible entry
   d. Dry-chemical extinguishing agents

____ 34. Which type of in-flight emergency is most likely to affect aircraft steering, braking, and/or stopping?
   a. Engine failure
   b. Onboard fire
   c. Malfunctioning flight controls
   d. Hydraulic problems

____ 35. All of the following statements about in-flight fires are true *except* _____.
   a. Interior fires are usually *not* detected until their development stage.
   b. Most in-flight interior fires are *not* true emergencies.
   c. Any time an in-flight fire is discovered, an emergency landing is attempted immediately.
   d. After the aircraft lands, rescue workers must vent the aircraft as quickly as possible.

____ 36. Which of the following statements about aircraft emergency evacuation assistance is true?
   a. Occupants who *cannot* extricate themselves should be evacuated first.
   b. Firefighters should discharge agent onto the fuselage of an aircraft with an interior fire.
   c. The most important aspect is opening the aircraft.
   d. ARFF crew members should position themselves at the front of escape slides to assist occupants.

____ 37. Low-impact crashes are those that _____.
   a. Occur at ground speeds less the 15 mph *(24 kmph)*
   b. Do *not* result in serious injuries
   c. Do *not* require rescue operations
   d. Do *not* severely damage or break up the fuselage

10

____ 38. Which of the following statements about wheels-up landings is true?
   a. Fire is uncommon.
   b. They are least hazardous when the plane lands on an airport runway.
   c. ARFF vehicles should begin their pursuit as soon as the aircraft touches down.
   d. It is almost impossible for the pilot to maintain control of the aircraft.

____ 39. A low-impact wheels-up landing on water is known as _____.
   a. Ditching
   b. Splashdown
   c. Flotation
   d. Treading water

____ 40. Which of the following statements about low-impact landings on water is true?
   a. If practical, personnel should contain fuel spills with floating beams.
   b. Making an opening above the water level may cause the wreckage to submerge.
   c. Neoprene wet suits offer better protection than dry suits in cold temperatures.
   d. For aircraft accidents in swamps, marshes, and tidal flats that are inaccessible by conventional rescue boats and land vehicles, helicopters are the best alternative.

____ 41. The main wreckage from a helicopter crash usually is the _____.
   a. Interior of the fuselage
   b. Rotor
   c. Undercarriage
   d. Tail unit

____ 42. Which of the following is *not* by definition a high-impact crash?
   a. Badly damaged fuselage
   b. G forces upon occupants exceeding human tolerance
   c. Overrun of runway resulting from sudden loss of power
   d. Failure of seats and safety devices to restrain passengers

____ 43. All of the following are causes of aircraft controlled flight into terrain *except* _____.
   a. Mechanical problem
   b. Weather conditions
   c. Pilot distraction
   d. Improper instrument settings

____ 44. The minimum information that ARFF units responding to an aircraft accident should have includes all of the following *except* _____.
   a. Make and model of aircraft
   b. Type and location of onboard extinguishing systems
   c. Number and locations of occupants
   d. Amount of fuel on board

45. If an ARFF apparatus driver/operator's vision becomes obscured during an emergency response and two people are aboard, the second person should _____.
    a. Attempt to eliminate the problem.
    b. Stand on the front bumper and guide the driver/operator.
    c. Sweep the area in front of the vehicle.
    d. Walk alongside the vehicle and guide the driver/operator.

46. Additional apparatus must be able to respond to an aircraft emergency and begin extinguishment within _____ minutes.
    a. Two
    b. Three
    c. Four
    d. Five

47. The main objective during the initial attack at an aircraft emergency is _____.
    a. Rescue
    b. Extinguishment
    c. Containment
    d. Scene control

48. How should turrets and ground sweeps be used at an aircraft emergency?
    a. To back up handlines
    b. To mount an interior fire attack
    c. To facilitate overhaul
    d. To control fire around the exterior fuselage

49. When conditions at an exterior aircraft fire permit, where should ARFF apparatus be positioned?
    a. On both sides of the aircraft
    b. At the nose or one side of the aircraft
    c. At the tail or one side of the aircraft
    d. At the nose or tail of the aircraft

50. Personnel working in the aircraft interior during ventilation operations should have a charged hoseline with a _____ nozzle.
    a. Straight stream
    b. Combination
    c. Broken stream
    d. Fog

51. Which of the following methods generally provides the fastest means of aircraft evacuation?
    a. Steps
    b. Ladders
    c. Slides
    d. Ropes

52. What is the best way to disable an aircraft's electrical system during extinguishment activities?
    a. Rely on "power-off" switches.
    b. Disconnect the batteries' negative terminals and leave the positive terminals connected.
    c. Disconnect batteries and remove them.
    d. Disconnect batteries and tape their terminals.

10

___ 53. Which of the following statements about the extinguishment phase is true?
   a. The extinguishment phase and the rescue phase may be conducted simultaneously.
   b. The extinguishment phase is merely an extension of the initial attack.
   c. The extinguishment phase is the final phase.
   d. Crash debris should be moved to facilitate extinguishment.

___ 54. Which of the following activities at an aircraft crash requires authorization from the responsible investigator?
   a. Requesting additional water-supply vehicles
   b. Using special lighting and air-supply units
   c. Dispatching special purpose vehicles that carry mass quantities of medical supplies
   d. Using wreckers to move parts of the wreckage

___ 55. Which of the following statements about aircraft overhaul is true?
   a. Overhaul is the least dangerous phase of aircraft fire fighting.
   b. ARFF personnel may remove SCBA as soon as overhaul begins.
   c. ARFF personnel must leave all wall panels, partitions, and ceiling coverings in place.
   d. Visible smoke and/or steam usually indicate the location of hot spots.

___ 56. Which of the following documents furnishes general guidance for preservation of evidence at aircraft accidents?
   a. FAA AC 150/5200-12B   c. NTSB 1978.605
   b. NFPA 403   d. NFPA 1003

___ 57. When ARFF personnel must move a body, they should tag it and note _____.
   a. A description of the remains
   b. Where the body was found
   c. Names of personnel who moved the remains
   d. Probable cause of death

___ 58. Substances that ignite spontaneously on contact with each other are called _____.
   a. Hypergolics   c. Hypertonics
   b. Hyperploids   d. Hyperboreans

___ 59. Military aircraft may have emergency power units that use hydrazine, an explosive and highly toxic substance with an odor similar to _____.
   a. Chlorine   c. Ammonia
   b. Fluoride   d. Sulfur

10

___ 60. Which of the following statements about aircraft accident victim management is true?
   a. A high-impact crash will almost certainly require personnel to address emergency medical services.
   b. Low-impact crashes usually require no more than routine medical services for victims.
   c. When treating victims, personnel should first ensure personal protection against bloodborne pathogens.
   d. A yellow triage tag indicates low priority or walking wounded.

___ 61. Once triaged and tagged with the level of urgency, victims should be _____.
   a. Loaded on a transport vehicle
   b. Moved to a treatment area
   c. Treated where they were found
   d. Monitored until more qualified personnel arrive

___ 62. The document that outlines the resources needed for an aircraft crash is called a(n) _____.
   a. Crisis Action Plan   c. Airport Emergency Plan
   b. Crash Management Plan   d. Aircraft Disaster Plan

Identify

C. Identify guidelines for positioning ARFF apparatus. Write an *X* before each correct statement below.

___ 1. Drive through smoke as necessary to reach fleeing passengers.

___ 2. Always attempt to position vehicles downhill and downwind.

___ 3. Position apparatus so that they block the entry to the accident site.

___ 4. Position vehicles so that they may be operated quickly in the event of flash fire.

___ 5. Position vehicles to protect the egress route of persons from the aircraft.

___ 6. Position vehicles so that they can be rapidly backed into new positions.

___ 7. Position vehicles so that turrets and handlines *cannot* accidentally be directed at the route of egress.

103
Aircraft Rescue and Fire Fighting

10

D. Identify guidelines for responding to unignited fuel leaks or spills. Write an *X* before each correct statement below.

_____ 1. Attempt to shut off the fuel at the source.

_____ 2. Confine occupants to the aircraft.

_____ 3. Make sure that fire-fighting personnel are wearing full protective clothing, including SCBA.

_____ 4. If necessary, blanket all exposed fuel surfaces with foam.

_____ 5. Disperse spilled fuel over as wide an area as possible.

_____ 6. Direct spilled fuel toward runoffs.

_____ 7. Keep apparatus and equipment ready to protect rescue operations.

E. Identify guidelines for responding to fuel spills occurring as a result of a collision. Write an *X* before each correct statement below.

_____ 1. Evacuate the aircraft.

_____ 2. Do not allow anyone to walk through the liquid fuel.

_____ 3. Wipe fuel contamination from the skin with dry absorbent materials.

_____ 4. Do not start any spark-producing equipment in the area before the spilled fuel is blanketed or removed.

_____ 5. If a vehicle engine is running at the time of the spill, shut it down.

_____ 6. Increase engine speed before shutting down any vehicle with an internal combustion engine.

_____ 7. If any aircraft engine is operating at the time of the spill, shut it down.

Chapter 11 Airport Emergency Plans

True/False

A. Write *True* or *False* before each of the following statements. Correct those statements that are false.

_____ 1. Developing an airport emergency plan is an end in itself.

_____ 2. A document that offers information pertinent to airport emergency planning is FAR Part 139.325, *Airport Emergency Plan*.

_____ 3. In many cases, a single, detailed airport emergency plan can serve the needs of all agencies involved.

_____ 4. ARFF personnel should handle accidents/incidents involving general aviation aircraft with greater urgency than they do those involving commercial aircraft.

_____ 5. All concerned with aircraft emergency services must have grid maps of the airport and surrounding areas within a 5- to 10-mile *(8 km to 16 km)* radius.

_____ 6. The nature of the ground surface plays an important part in aircraft rescue operations and extinguishing methods.

105
Aircraft Rescue and Fire Fighting

_____ 7. A hangar or terminal building may be used to accommodate aircraft passengers and crew members with minor injuries.
_____
_____

_____ 8. Only emergency vehicles should be used to transport aircraft accident survivors who are injured but ambulatory.
_____
_____

_____ 9. The emphasis in patient care following most aircraft accidents is on immediate transport of the injured to the nearest medical facility.
_____
_____

_____ 10. The secondary response network is activated as the primary response units arrive on the scene of an aircraft accident/incident.
_____
_____

_____ 11. Airport emergency plans should include contractual information to ensure the availability of all necessary heavy equipment and special-purpose equipment.
_____
_____

_____ 12. Authorities must withhold the names of those not seriously injured until any fatalities' next of kin have been notified.
_____
_____

_____ 13. Public law in the United States allows military agencies to enter into reciprocal mutual-assistance agreements with surrounding community fire protection organizations.
_____
_____

_____ 14. Refrigerated trucks may be used as temporary morgues when aircraft fatalities overwhelm local morgue facilities.

_____ 15. Military police have peace-officer authority at all accidents/incidents involving military aircraft.

Multiple Choice

B. **Write the letter of the best answer on the blank before each statement.**

_____ 1. The best way to train personnel for executing an airport emergency plan is to begin with _____.
   a. A large-scale airport disaster simulation
   b. Lectures giving theoretical background
   c. Small exercises that focus on specific tasks
   d. A written test over the plan

_____ 2. Because of the many variables, all agencies involved in an airport emergency plan must recognize the need for validating their plan through _____.
   a. Joint training exercises
   b. An external audit
   c. Internal reviews
   d. The national certification panel

_____ 3. Sections that should be common to all airport emergency plans include all of the following *except* _____.
   a. Civil disturbances
   b. Introduction
   c. Hazardous materials emergencies
   d. Background

_____ 4. Which of the following statements about types of accidents/incidents is true?
   a. Most aircraft incidents create an immediate risk to the occupants of the aircraft.
   b. In a low impact crash, if egress is *not* blocked by fire, fatality rates tend to be low.
   c. The number one goal of ARFF personnel at aircraft incidents is fire control.
   d. A small percentage of aircraft accidents result in fire.

5. The greatest potential for aircraft accidents is during _____.
   a. Landing/takeoff
   b. Cruising
   c. Taxiing
   d. Banking maneuvers

6. A high percentage of all aircraft accidents occur _____.
   a. 50 miles *(80 km)* or more from the departure or destination airport
   b. In the threshold or departure area of the runway
   c. On taxiways or aprons
   d. At populated areas of airports

7. Personnel can determine where airport emergencies are likely to occur by studying the _____.
   a. Posted flight schedules and information
   b. Types of aircraft using the airport
   c. Arrival and departure traffic patterns
   d. Average level of training for pilots using the airport

8. Which of the following statements about climatic conditions during ARFF operations is true?
   a. Heavy vehicles and equipment designed for airport rescue and fire fighting are virtually unaffected by climatic considerations.
   b. Portable shelters should be erected as soon as possible at every crash site under adverse weather conditions.
   c. During unfavorable weather, a climate-controlled rehabilitation area should be within walking distance from the site of all crashes.
   d. The airport emergency plan should provide for protecting aircraft occupants along with all emergency responders from harsh weather conditions.

9. Which of the following can be used to initiate an automatic recall of all off-duty fire-fighting personnel?
   a. Mobile radio alert system
   b. Audible alarm system
   c. Automated dialing system
   d. Call forwarding override system

10. At the very least, the primary response to an aircraft emergency should include the _____.
    a. Air carrier/owner
    b. Airport management
    c. Mutual response agencies
    d. News media clearinghouse

11. Primary responsibilities for law enforcement personnel at the site of an aircraft accident/incident include all of the following *except* _____.
    a. Traffic and crowd control
    b. Large-scale evacuation
    c. Fire extinguishment assistance
    d. Cordoning off the area

____ 12. Which of the following statements about emergency medical supplies for aircraft accidents/incidents is true?
   a. All barrier tape should have black markings on an optic yellow background.
   b. Most airports have a cache of medical supplies in a designated trailer or vehicle that can be transported to the scene when needed.
   c. Triage tags and marking pens are used to mark body parts or other important evidence.
   d. Deceased victims should be placed in body bags as soon as they are located.

____ 13. What minimum level of medical care should a temporary field hospital at aircraft accidents/incidents be organized and equipped to provide?
   a. Triage
   b. Basic first aid
   c. Stabilization and maintenance
   d. Pain management and emergency surgery

____ 14. A field hospital unit should have at least enough supplies to handle an accident involving _____ that use nearby airports.
   a. One of the most common size aircraft
   b. Two of the most common size aircraft
   c. One of the largest aircraft
   d. Two of the largest aircraft

____ 15. Air carrier organizations and their personnel may typically provide all of the following *except* _____.
   a. Information regarding the number of occupants
   b. Family-assistance resource teams
   c. Auxiliary airport firefighter units
   d. Information regarding known quantities of hazardous materials

____ 16. All of the following statements about mutual aid support for aircraft accidents are true *except* _____.
   a. In order to coordinate the efforts of all those who may be involved, planners should meet with representatives of all entities as a group.
   b. ARFF personnel must contact a mutual aid dispatch center and formally request additional manpower and equipment from mutual aid services on the primary response notification list.
   c. Mutual aid plans should include a comprehensive list of the resources available from the various entities, along with the key emergency phone numbers.
   d. Mutual aid agreements and lists must be periodically reviewed and revised.

17. Which of the following statements about rehabilitation is true?
    a. Response personnel should be required to spend time in the rehab area.
    b. The rehab area should be as close as possible to the accident site.
    c. The rehab area should be combined with triage on large incidents.
    d. Each response agency should stock its own supplies for nourishment and rehydration.

18. The type of psychological pressure that requires critical incident stress debriefing is created in situations that involve ____.
    a. Unusual danger to the responder
    b. Multiple injuries or fatalities
    c. Extraordinary physical exertion
    d. Extensively damaged aircraft

19. When should critical incident stress debriefing begin for personnel involved in rescue operations at a major aircraft accident?
    a. During the incident
    b. Within 24 hours of the incident
    c. Within 48 hours of the incident
    d. Within 72 hours of the incident

20. Only an investigation by the ____ can determine whether a crime was involved in an aircraft accident/incident.
    a. Local law enforcement agencies
    b. Fire marshal
    c. Federal Aviation Administration or International Civil Aviation Organization
    d. National Transportation Safety Board or Canadian Transportation Accident Investigation and Safety Board

21. Which of the following consults with the families of victims to arrange nondenominational memorial services just days following an aircraft crash?
    a. Coroner/medical examiner
    b. Critical incident stress debriefing team
    c. American Red Cross
    d. National Transportation Safety Board/Canadian Transportation Accident Investigation and Safety Board

22. Which of the following statements regarding communications for multiagency operations is true?
    a. All agencies should have at least one common channel for day-to-day operations.
    b. Each agency should have multichannel scanning capabilities.
    c. All agencies should use a mutually agreed upon radio code.
    d. The Incident Management System discourages the use of Clear Text.

23. Which of the following agencies should be notified that an accident involving military aircraft has occurred?
    a. FEMA
    b. FAA
    c. NTSB
    d. NFPA

___ 24. When responding to military aircraft accidents/incidents, the ranking officer from the base fire department may do any of the following *except* ___.
   a. Become part of a unified command team.
   b. Act as technical adviser to the IC.
   c. Assume command.
   d. Reassign command to another civilian authority.

___ 25. Which agency investigates military aircraft accidents/incidents that do *not* also involve civilian aircraft?
   a. National Transportation Safety Board
   b. Federal Bureau of Investigation
   c. Military Accident Investigation Board
   d. Central Intelligence Agency

___ 26. Who is responsible for removing the wreckage of any civilian aircraft involved in an accident/incident with military aircraft?
   a. Military personnel
   b. ARFF personnel
   c. Owners of the civilian aircraft
   d. Airport management

___ 27. Joint training exercises among participating agencies should emphasize ___.
   a. Overall strategies
   b. Individual fire-fighting skills
   c. Tactical considerations
   d. Management techniques

___ 28. How often should an entire airport emergency plan be practiced by conducting at least a tabletop exercise?
   a. Every month
   b. Every three months
   c. Every six months
   d. Every twelve months

Identify

C. Identify the following abbreviations associated with airport emergency plans. Write the correct interpretation before each.

_____ 1. AEP

_____ 2. NTSB

_____ 3. CTAISB

_____ 4. PIO

_____ 5. FEMA

_____ 6. NDA

_____ 7. EOD

11

D. Identify guidelines for dealing with news media personnel. Write an *X* before each correct statement below.

___ 1. Emergency personnel should avoid relationships with news media personnel.

___ 2. Emergency personnel should periodically meet with news media representatives to discuss mutual concerns.

___ 3. At an incident, emergency personnel should refer media representatives to the public information officer.

___ 4. Emergency personnel should not allow media representatives to remain at the scene of the accident.

___ 5. Emergency personnel should cooperate fully with media personnel as long as it does not interfere with rescue or fire-fighting efforts.

___ 6. The PIO is responsible for the release of information regarding commercial or military aircraft accidents/incidents.

___ 7. The law permits photographs to be taken of anything at the scene of a civil aircraft accident as long as no physical evidence is disturbed in the process.

E. Identify guidelines for training mutual aid and support personnel. Write an *X* before each correct statement below.

___ 1. All parties to mutual aid agreements should participate in airport emergency planning, training, and drills.

___ 2. Fire department response times are equally critical in aircraft accidents and structural fires.

___ 3. A thorough and effective airport emergency plan is sufficient to ensure that mutual aid personnel can perform their duties quickly and efficiently.

___ 4. Joint training exercises, drills, and tests should be conducted at the airport.

___ 5. Mutual aid companies should know airport terminology and control-tower light signals.

___ 6. Mutual aid companies should be trained to combat aircraft fires alone if necessary.

___ 7. Training structural firefighters to combat aircraft fires requires greater emphasis than training ARFF personnel to combat structural fires.

___ 8. Classes for all airport employees should acquaint them with the use of fire extinguishers, fire reporting procedures, and evacuation procedures.

Chapter 12   Hazards Associated with Aircraft Cargo

Matching

A. Match to their U.S. DOT classifications the following types of dangerous goods. Write the appropriate letters on the blanks.

____ 1. Flammable liquids

____ 2. Oxidizing substances

____ 3. Corrosives

____ 4. Radioactive materials

____ 5. Explosives

____ 6. Flammable solids

____ 7. Poisonous and infectious substances

____ 8. Miscellaneous dangerous goods

____ 9. Gases

a. Class 1
b. Class 2
c. Class 3
d. Class 4
e. Class 5
f. Class 6
g. Class 7
h. Class 8
i. Class 9
j. Class 10

B. Match to their U.S. EPA classifications the following protective clothing requirements. Write the appropriate letters on the blanks.

____ 1. Liquid/airborne particulate protection; nonencapsulating suits

____ 2. No specific protection; nonencapsulating suits

____ 3. Liquid protection; nonencapsulating suits

____ 4. Vapor/gas protection; totally encapsulating suits

a. Level A
b. Level B
c. Level C
d. Level D
e. Level E

12

True/False

C. Write *True* or *False* before each of the following statements. Correct those statements that are false.

_____ 1. The term *hazardous materials* is used in both the United States aviation industry and fire service.

_____ 2. Almost all dangerous goods except for Class A explosives and poisonous gases can be shipped by air.

_____ 3. Requirements that air carriers must inspect packages and documents prepared by the shipper guarantee that only proper shipments are on board an aircraft.

_____ 4. Certain hazardous cargoes such as dry ice must always be stowed in specialized containers.

_____ 5. Properly marked dangerous goods will always be easily identified.

_____ 6. The means of identifying dangerous goods include the UN/NA number.

_____ 7. A mitigation plan should be devised as soon as any dangerous goods problem is identified.

8. In situations involving unknown materials, the role of ARFF personnel may be limited to isolating the contaminated area and denying entry until a hazardous materials response team can obtain a sample of the material for analysis.

9. The IC has authority to decide that the potential benefit outweighs the risk of subjecting firefighters wearing SFPC to a "quick in-and-out" mission of critical importance.

10. When large-scale evacuations become necessary at a dangerous good incident, they probably will be the responsibility of ARFF personnel.

11. ARFF personnel may be put at moderate risk at a dangerous goods incident to effect rescues and recover bodies.

Multiple Choice

D. Write the letter of the best answer on the blank before each statement.

1. All of the following statements about dangerous goods are true *except* ____.
   a. Emergency response to aircraft carrying dangerous goods usually remains unchanged from normal aircraft fire-fighting procedures.
   b. Any aircraft crash could be considered to involve dangerous goods, regardless of the cargo.
   c. Responders may safely handle small amounts of the composites used in aircraft design and construction without using personal protective equipment.
   d. Any aircraft subjected to the dynamics of a crash and subsequent fire may release harmful substances.

2. Dangerous goods shipments by civilian aircraft are regulated by ____.
   a. CFR Title 49, Part 175
   b. NFPA 402
   c. ICAO *Airport Services Manual*, Part 2, Chapter 9
   d. NIOSH 613-A

12

____ 3. Which type of border on shipping papers should alert ARFF personnel to the presence of dangerous goods on an aircraft?
   a. Solid black
   b. Solid red
   c. Black and yellow stripe
   d. Red and white stripe

____ 4. A dangerous goods warning label that has a solid blue background indicates ____.
   a. Poison gases
   b. Oxidizers
   c. Dangerous when wet materials
   d. Infectious substances

____ 5. A dangerous goods warning label whose top half is white and bottom half is black indicates ____.
   a. Explosives
   b. Corrosives
   c. Miscellaneous dangerous goods
   d. Organic peroxides

____ 6. What color(s) are the backgrounds of dangerous goods warning labels for radioactive materials?
   a. Solid yellow
   b. Red and white
   c. Yellow and white
   d. Solid orange

____ 7. On cargo-carrying aircraft, hazardous freight is usually placed in containers called ____.
   a. Unit load devices
   b. Single compartment boxes
   c. Dangerous good bins
   d. Haz mat repositories

____ 8. Which type of portable extinguishers is used with containers that have an integral fire suppression capability?
   a. Water
   b. Foam
   c. Halon 1211
   d. Dry chemical

____ 9. As a general rule, most dangerous goods on the main deck of cargo aircraft are loaded ____.
   a. In the most forward location
   b. Directly over the wings
   c. As far away from the flight crew as possible
   d. Wherever they can be accessed by a path or aisle

____ 10. As a general rule, radioactive materials on the main deck of cargo aircraft are loaded ____.
   a. In the most forward location
   b. Directly over the wings
   c. As far away from the flight crew as possible
   d. Wherever they can be accessed by a path or aisle

11. Cargo that is *not* packaged properly, does *not* have shipping documentation, or has *not* been handled with the required safety precautions is considered _____ dangerous goods cargo.
    a. Undisclosed
    b. Unprotected
    c. Unexposed
    d. Undeclared

12. On passenger aircraft, where are hazardous goods shipping papers maintained?
    a. In the cargo bay with the hazardous goods
    b. On the flight deck
    c. In a pouch near an exit door
    d. In the cargo bay just aft of the bay containing the dangerous goods

13. At least how many separate sources should be used to identify and verify hazardous goods?
    a. Two
    b. Three
    c. Four
    d. Five

14. All of the following are common sources of information about hazardous goods *except* _____.
    a. *Hazardous Materials Containment and Cleanup Guidebook*
    b. *Fire Protection Guide on Hazardous Materials*
    c. *Dangerous Goods Initial Emergency Response Guide*
    d. *2000 Emergency Response Guidebook*

15. Who determines the type of personal protective equipment to be used during responses involving dangerous goods?
    a. Individual firefighters
    b. Local OSHA administrators
    c. Authorities having jurisdiction
    d. Company officers

16. The three basic levels of protection for the emergency responder include all of the following *except* _____.
    a. Structural fire fighting clothing
    b. Proximity clothing
    c. Chemical protective clothing
    d. Approach clothing

17. Structural fire fighting protective clothing meets the requirements for EPA _____.
    a. Level A
    b. Level B
    c. Level C
    d. Level D

18. The first responsibility of units responding to dangerous goods incidents is to _____.
    a. Identify and verify the materials.
    b. Isolate the scene and deny entry.
    c. Evaluate the scene and mitigate the hazards.
    d. Establish a support zone and stay there.

12

____ 19. Which of the following statements about aircraft incidents involving dangerous goods is true?

   a. Aircraft design and construction inherently use so many hazardous materials that cargoes of dangerous goods have little significant effect on the risk factors for ARFF personnel.
   b. The dangerous goods that could be found on or around aircraft are the same as those that could be found on the highway or in a fixed facility.
   c. The quantities of hazardous goods involved in most aircraft accidents tend to be greater than those in other environments.
   d. Exposures to minute amounts of hazardous materials involve minimal risks to personnel.

____ 20. How are agricultural chemicals usually stored at the aircraft loading point?

   a. As liquids in storage tanks and as powders in silos
   b. As both liquids and powders in 1-gallon *(4 L)* plastic cubes
   c. As liquids in underground holding tanks and as powders in bales
   d. As liquids in storage drums and as powders in bags

____ 21. The quantities of chemicals carried in agricultural spraying or crop dusting aircraft can range up to ____.

   a. Several thousand gallons *(liters)*
   b. About one thousand gallons *(liters)*
   c. Several hundred gallons *(liters)*
   d. About one hundred gallons *(liters)*

____ 22. Which of the following is usually the most reliable indicator that an aircraft was used for spraying/crop dusting?

   a. Application equipment in the wreckage
   b. Hazardous materials warning placards on the aircraft
   c. Material safety data sheets in a pouch near the main exit door
   d. Skull and crossbones poison symbol on the underside of the fuselage

Identify

E. Identify the following abbreviations associated with hazards associated with aircraft cargo. Write the correct interpretation before each.

_____ 1. DG

_____ 2. IATA

_____ 3. RQ

_____ 4. CHRIS

_____ 5. BOE

_____ 6. AAR

118
Aircraft Rescue and Fire Fighting

12

_____ 7. CHEMTREC

_____ 8. CANUTEC

_____ 9. HMRT

_____ 10. EPA

_____ 11. SFPC

Chapter 1 Answers

True/False

A.
1. False. ACs *are not requirements* but rather *suggested or recommended guidelines*. *(3–4)*
2. True *(4)*
3. True *(4)*
4. False. Classroom training and looking at maps *do not provide* all the information a firefighter needs to become familiar with an airport. The firefighter *must get out on the airport grounds* and learn the airport features. *(5)*
5. True *(6)*
6. False. The need for security at an airport is the responsibility of *all airport employees*. *(6)*
7. True *(9)*
8. False. ARFF vehicles are designed to deliver mass quantities of water/agent, but *they have limited capacities*. Therefore, agent management *is important* to successful ARFF operations. *(9)*
9. False. Most apparatus or equipment failures *can be prevented* by performing routine maintenance checks on a regular basis. *(11)*
10. False. Depending on the material burning and the size and location of the fire, different situations may require *a different type of extinguishing agent*. *(11)*
11. True *(11)*
12. True *(11)*
13. False. Aircraft cargo *may be inherently hazardous*, or it may become hazardous as a result of fire. *(12)*
14. True *(12)*
15. True *(12)*

Multiple Choice

B.
1. b *(3)*
2. a *(4)*
3. b *(4–5)*
4. c *(5)*
5. d *(5)*
6. a *(5)*
7. d *(5)*
8. c *(5)*
9. b *(6)*
10. d *(8)*
11. a *(9)*
12. a *(9)*
13. d *(10)*
14. c *(10, 11)*
15. a *(10)*
16. c *(11)*
17. b *(13)*
18. b *(13)*

Identify

C.
1. Aircraft rescue and fire fighting *(3)*
2. National Fire Protection Association *(3)*
3. *Code of Federal Regulations* *(3)*
4. Federal Aviation Regulation *(3)*
5. International Civil Aviation Organization *(3)*
6. Federal Aviation Administration *(3–4)*
7. Advisory Circular *(4)*
8. Auxiliary power unit *(8)*
9. Digital flight data recorder *(8)*
10. Cockpit voice recorder *(8)*
11. Personal protective equipment *(8)*
12. Incident Management System *(8)*
13. Critical incident stress debriefing *(8)*
14. *Emergency Response Guidebook* *(13)*
15. Chemical Transportation Emergency Center *(13)*
16. Canadian Transport Emergency Centre *(13)*

Chapter 2 Answers

Matching

A.
1. c *(18)*
2. f *(18)*
3. a *(18)*
4. e *(18)*
5. j *(18)*
6. i *(20)*
7. b *(23)*
8. d *(28)*
9. g *(29)*

True/False

B.
1. True *(15)*
2. True *(16)*
3. False. ARFF vehicles *may use* the same access routes as aircraft, so it is important that the airport firefighter know the meanings of runway and taxiway designation systems. *(18)*
4. False. Taxiway designations *are not standardized* and are generally *determined locally*. *(18)*
5. True *(20)*
6. True *(22)*
7. False. The CRFFAA's *width* extends 500 feet *(152 m)* from each side of the runway centerline, and its *length is 3,300 feet (1 000 m)* beyond each runway end. *(22)*
8. True *(23–24)*
9. False. Airport firefighters *do not need to know* the details of how navigational aids work. They should, however, be able to *identify navigation aids and know their locations* on the airport. *(26)*
10. True *(26)*
11. True *(26)*
12. False. When feasible, airport firefighters *should visually monitor* current weather conditions. *(28)*
13. True *(29)*
14. False. Distribution of water from fixed water supply systems *is through domestic water supply mains* unless the system is designed otherwise. *(29)*
15. True *(29)*
16. False. Grounding to a static ground electrode in the pavement during fueling operations *is not required* by NFPA standards but may still be requested by the carrier or required by other standards. *(30)*
17. False. Open flames should be strictly controlled or prohibited *within 50 feet (15 m)* of any aircraft fueling operation. *(32)*
18. True *(32)*
19. False. Multipurpose dry chemical fire extinguishers *are no longer recommended* for use during aircraft fueling operations because of their *corrosive effects* on aircraft materials. *(32)*
20. False. *Not all airports* have the same type of drainage system, and firefighters must know the type of drainage system design at their airports. *(32)*

Multiple Choice

C.
1. b *(16)*
2. d *(16)*
3. d *(16)*
4. a *(16)*
5. b *(16)*
6. c *(17)*
7. a *(18)*
8. d *(18)*
9. d *(18)*
10. d *(19)*
11. b *(19)*
12. a *(19)*
13. b *(20)*
14. c *(20)*
15. c *(20)*
16. c *(20)*
17. a *(20)*
18. d *(21)*
19. b *(21)*
20. b *(21)*
21. a *(21)*
22. c *(21)*
23. d *(22)*
24. c *(22)*
25. b *(22)*
26. d *(23)*
27. a *(24)*
28. b *(24)*
29. d *(25)*
30. c *(25)*
31. a *(25)*
32. c *(26)*
33. b *(26)*
34. a *(28)*
35. d *(28)*
36. d *(30)*
37. b *(30)*
38. b *(30)*
39. c *(31)*
40. b *(31)*
41. d *(31)*
42. a *(31)*
43. a *(32)*
44. b *(32)*
45. c *(32)*

Identify

D.
1. Authority having jurisdiction *(16)*
2. Instrument Landing System *(20)*
3. Air traffic control *(20)*
4. Critical rescue and fire fighting access area *(22)*
5. Navigational aid *(26)*
6. Ground power unit *(27)*
7. Foreign object debris *(28)*
8. Security identification display area *(28)*

Chapter 3 Answers

Matching

A.
1. e *(40)*
2. l *(40)*
3. h *(41)*
4. c *(41)*
5. k *(41)*
6. d *(41)*
7. m *(41–42)*
8. a *(42)*
9. i *(42)*
10. g *(42)*
11. j *(42)*
12. b *(42)*

True/False

B.
1. True *(34)*
2. True *(37)*
3. False. Each type of military aircraft is identified by a letter prefix that indicates its *function;* the A-10 is an *attack* aircraft. *(38)*
4. True *(39)*
5. False. The internal fuel tanks in rotary wing aircraft are usually located *under the cargo floor.* *(39)*
6. False. Conventional landing gear consists of a *tail wheel* and two main struts under each wing. *(40)*
7. True *(42)*
8. True *(42)*
9. False. Disconnecting the battery *does not prevent* the magneto from functioning, so unspent fuel in the engine cylinders may be ignited, causing the engine to restart and the propeller to rotate. *(43)*
10. True *(48)*
11. False. When a number of jet engines are operating in an area, it is *often difficult* for ground personnel to tell which engines are operating and which are not, especially if the personnel are wearing hearing protection. *(48)*
12. True *(49)*
13. False. The FAA *does not* consider stowage compartments such as overhead storage areas for carry-on articles and baggage as cargo compartments. *(49)*
14. False. Class D compartments are *no longer an option* for new aircraft. *(50)*
15. True *(50)*
16. False. Pneumatic drivers *cannot be used* because they turn too fast and jam the mechanism. *(50)*
17. True *(50)*
18. False. One of aluminum's *disadvantages* for aircraft construction is that *it does not* withstand high heat well. *(50–51)*
19. True *(51)*
20. True *(54)*
21. False. Although damage to an aircraft may seem insignificant, *even minor damage can be critical* because leaking or seeping fuel may pool in low-lying sections of the fuselage. *(54–55)*
22. False. Although the technology for fuel tanks has continued to advance, these improvements *have not been widely adopted.* *(55)*
23. True *(55)*
24. True *(55)*
25. True *(57)*
26. False. Brakes and wheels may reach their maximum temperatures *20 to 30 minutes after* the aircraft comes to a stop. *(57)*
27. True *(58)*
28. False. All aircraft shutdown functions must be completed *prior to* de-energizing the electrical system. *(59)*
29. False. GPUs can be used to produce either AC or DC power and come in *diesel- or gas-fueled models.* *(60)*
30. True *(61)*
31. True *(62)*
32. False. One acceptable way to stop a liquid oxygen leak is to spray the leak with *water fog.* *(62)*
33. True *(62)*
34. False. Oxygen cylinders have pressure-relief valves, but cylinders used for hydraulic fluids, fire extinguishing systems, rain repellent, and pneumatic systems *may explode* during aircraft fire-fighting operations. *(63)*
35. True *(64)*
36. True *(67)*
37. False. The "black boxes" are actually painted *international orange* or *bright red.* *(69)*
38. False. If a military aircraft functions in a dual role, it carries the letter for which it was *originally designed.* *(70)*
39. True *(72)*
40. False. The operation of the exit doors and hatches *varies with each model* of aircraft, and some familiarity with the aircraft is necessary to effectively operate them under emergency conditions. *(73)*

3

41. False. Ejection systems are normally found in fighter, attack, bomber, *and training aircraft*. *(73)*
42. True *(74)*
43. False. Aircraft canopies weigh *several hundred pounds (kilograms)*. *(74)*
44. True *(76)*
45. False. If a cast high explosive melts and runs from a ruptured ammunition case, it becomes *extremely sensitive to shock after* it resolidifies. *(77)*
46. True *(78)*

Multiple Choice

C.

1. c *(34)*	39. b *(49)*	77. d *(66)*
2. c *(35)*	40. d *(50)*	78. c *(66)*
3. a *(35)*	41. a *(49)*	79. a *(66)*
4. d *(35)*	42. d *(50)*	80. c *(66)*
5. c *(35–36)*	43. c *(51)*	81. b *(67)*
6. d *(36)*	44. d *(51)*	82. a *(67)*
7. b *(36–37)*	45. a *(51)*	83. d *(68)*
8. b *(37)*	46. d *(52)*	84. b *(68)*
9. c *(37)*	47. b *(53)*	85. d *(68)*
10. c *(37, 38)*	48. c *(53)*	86. c *(69)*
11. d *(38)*	49. b *(54)*	87. d *(69)*
12. c *(39)*	50. a *(55)*	88. c *(70)*
13. a *(39)*	51. c *(55)*	89. b *(70)*
14. b *(41)*	52. d *(56)*	90. a *(70)*
15. a *(41)*	53. d *(56)*	91. a *(70)*
16. d *(41)*	54. b *(56)*	92. d *(71)*
17. c *(43)*	55. b *(56)*	93. c *(71)*
18. b *(43)*	56. a *(57)*	94. b *(71)*
19. b *(43)*	57. d *(57)*	95. b *(71–72)*
20. d *(43)*	58. c *(58)*	96. c *(72)*
21. a *(44)*	59. c *(59)*	97. a *(72)*
22. a *(44)*	60. b *(59)*	98. d *(73)*
23. c *(44)*	61. d *(59)*	99. a *(75)*
24. d *(44)*	62. a *(60)*	100. d *(75)*
25. b *(45)*	63. b *(60)*	101. c *(75)*
26. c *(45)*	64. a *(60)*	102. b *(75)*
27. b *(45)*	65. c *(61)*	103. a *(75)*
28. c *(45)*	66. b *(61)*	104. c *(76)*
29. d *(45)*	67. d *(62)*	105. a *(76)*
30. a *(45–46)*	68. d *(62)*	106. b *(76)*
31. b *(46)*	69. b *(62)*	107. d *(76)*
32. b *(46)*	70. a *(63)*	108. c *(77)*
33. c *(46)*	71. c *(63)*	109. b *(77)*
34. a *(48)*	72. a *(63)*	110. a *(77)*
35. a *(48)*	73. d *(63)*	111. c *(78)*
36. d *(49)*	74. b *(64)*	112. d *(78)*
37. c *(49)*	75. b *(64, 65)*	113. c *(78)*
38. d *(49)*	76. c *(64–65)*	

Identify

D.

1. New large aircraft *(36)*
2. Very large aircraft *(36)*
3. Aviation gasoline *(38)*
4. Emergency power unit *(60)*
5. Ram-air-turbine *(60)*
6. Liquid oxygen *(62)*
7. Flight data recorder *(69)*
8. Cockpit voice recorder *(69)*
9. National Transportation Safety Board (U.S.) *(70)*
10. Jet-assisted takeoff *(71)*
11. High explosive *(77)*
12. Explosive ordnance disposal *(77)*

E. *(46, 47)*

1, 4

Chapter 4 Answers

True/False

A.
1. True *(80)*
2. False. Wearing a PASS device increases a firefighter's chances of being found in an emergency, but *only if it is turned on and working properly.* *(81)*
3. True *(81)*
4. False. *Supervisors are responsible* for knowing and enforcing the rules for vision protection without exception. *(82)*
5. False. Some activities such as operating power tools that generate sparks or clouds of dust and debris may require that the firefighter *also wear goggles or self-contained breathing apparatus.* *(82)*
6. True *(82)*
7. True *(83)*
8. True *(85)*
9. False. There will *always be the need* for personnel decontamination at a crash site. *(86)*
10. True *(86)*

Multiple Choice

B.
1. b *(80)*
2. a *(80)*
3. d *(80)*
4. c *(80)*
5. b *(81)*
6. b *(81)*
7. c *(81)*
8. a *(82)*
9. d *(82)*
10. d *(82)*
11. b *(82)*
12. c *(83)*
13. b *(83)*
14. a *(84)*
15. a *(84)*
16. d *(85)*
17. c *(86)*
18. d *(86–87)*
19. c *(87)*
20. b *(87)*

Identify

C.
1. Personal alert safety system *(80)*
2. Incident Management System *(80)*
3. Critical incident stress debriefing *(86)*
4. Material safety data sheet *(87)*

D. *(85–86)*
1, 4, 6, 7, 9, 10

125
Aircraft Rescue and Fire Fighting

Chapter 5 Answers

True/False

A.
1. False. Because other fire and law enforcement agencies and the local media *monitor public safety frequencies*, the manner in which communications are handled *projects an image* of the fire department. *(89–90)*
2. True *(90)*
3. False. The method used to notify firefighters of an aircraft incident/accident *may vary* from airport to airport or at military installations. *(90)*
4. False. Response categories *are not* standardized. *(90–91)*
5. True *(91)*
6. True *(93)*
7. True *(93)*
8. False. *Control tower personnel* direct aircraft and vehicular traffic by two-way radio on the ground-control frequency. *(97)*
9. False. Control tower personnel direct aircraft and vehicular traffic with light signals *if radio communication is lost.* *(97)*

Multiple Choice

B.

1. b *(90)*	8. b *(92)*	15. d *(93)*
2. c *(90)*	9. d *(92)*	16. c *(93)*
3. a *(90)*	10. d *(92)*	17. b *(94)*
4. c *(91)*	11. b *(92)*	18. d *(94)*
5. d *(91)*	12. c *(92–93)*	19. a *(98)*
6. b *(91)*	13. a *(93)*	20. d *(98)*
7. a *(92)*	14. c *(93)*	21. b *(98)*

Identify

C.
1. Notice to airmen *(90)*
2. Air traffic control tower *(92)*
3. Flight Service Station *(93)*
4. Visual flight rules *(93)*
5. Unified communications *(93)*
6. Common traffic advisory frequency *(93)*
7. Automated Terminal Information Service *(93)*
8. Mobile data terminal *(97)*
9. Global positioning system *(97)*

D. *(93)*
- A. 2
- B. 5
- C. 4
- D. 1
- E. 3

E. *(94)*
4

F. *(94, 95)*

Alpha	November
Bravo	Oscar
Charlie	Papa
Delta	Quebec
Echo	Romeo
Foxtrot	Sierra
Golf	Tango
Hotel	Uniform
India	Victor
Juliett	Whiskey
Kilo	X-ray
Lima	Yankee
Mike	Zulu

G. *(95–97)*
1. Acknowledge
2. Correction
3. Flameout
4. Hold your position
5. Low approach
6. Mayday
7. Out
8. Over
9. Roger
10. Stand by to copy
11. Verify
12. Wilco

H. *(97)*
1. Clear to cross, proceed, or go
2. Stop
3. Clear the taxiway/runway
4. Return to the starting point on the airport
5. Exercise extreme caution

I. *(99)*
1. Sound all audible devices for obviously extended time.
2. Flash headlights and sound siren.
3. Tap hand firmly on the desired nozzle barrel.
4. Place wrists together and clap hands.
5. Tug coattail sharply, or with the hands in front of the chest, give series of pushing motions.

Chapter 6 Answers

True/False

A.
1. False. Aviation industry standards require that ARFF apparatus be able to reach the scene of an aircraft emergency in *much shorter time periods* than those normally associated with a municipal fire department's response to a structure fire. *(101)*
2. True *(104)*
3. False. Batch mixing is suitable *only for regular AFFF* foam concentrates, *not alcohol-resistant* concentrates. *(104–105)*
4. False. These apparatus are intended to meet specific needs for particular airports; however, they generally *do not count* toward the minimum index requirements for fire protection. *(105)*
5. True *(105)*
6. True *(106)*
7. False. Both types of handlines on ARFF apparatus must be equipped with *variable pattern, shutoff nozzles*. Nozzles may be *aspirating or nonaspirating*. *(107)*
8. False. Because extendable turrets are mounted to the tops of ARFF vehicles, the vehicle may be *more top-heavy than other ARFF vehicles*. *(108)*
9. True *(108)*
10. True *(108)*
11. True *(109)*
12. False. The least desired method of resupplying an apparatus with foam is *direct filling from 5-gallon (20 L) containers*. *(109)*

Multiple Choice

B.
1. c *(102)*
2. a *(102)*
3. d *(102)*
4. b *(103)*
5. b *(103)*
6. c *(103)*
7. d *(103)*
8. c *(104)*
9. a *(104)*
10. d *(105)*
11. b *(106)*
12. c *(106)*
13. d *(106–107)*
14. c *(107)*
15. c *(108)*
16. d *(108)*
17. b *(108)*
18. b *(108)*
19. c *(108)*
20. a *(109)*
21. a *(109)*
22. c *(109)*
23. d *(110)*

Identify

C.
1. Driver's enhanced vision system *(106)*
2. Differential global positioning system *(106)*
3. Forward Looking Infrared *(108)*

Chapter 7 Answers

True/False

A.
1. False. In most cases, conventional fire-fighting tools *are* adaptable to aircraft rescue and forcible entry. *(112)*
2. True *(112)*
3. True *(113)*
4. True *(113)*
5. False. Many techniques that work in auto extrication *do not work as effectively* in the aircraft environment. *(113)*
6. True *(114)*
7. False. While pneumo-hydraulic units are *safe to use in flammable atmospheres*, they are not widely used. *(115)*
8. True *(115)*
9. False. The hydraulic pressure can be produced either *manually through a hand pump* or through a power unit. *(115–116)*
10. False. An *advantage* of using hydraulic tools in flammable areas is that they *do not produce sparks*. *(116)*
11. True *(117)*
12. False. The applications of rope in the ARFF environment *are the same* as for other operations; they include pulling, stabilizing, moving of tools and equipment, and creating barriers. *(117)*
13. True *(117, 118)*
14. True *(118)*
15. True *(118)*
16. False. If electrical systems fail, *the aircraft may have backup pneumatic systems that can operate the doors*. *(121)*
17. True *(121)*
18. False. Cuts should sever the *fewest possible* reinforcing channels, stiffeners, ribs, or longerons. *(122)*

Multiple Choice

B.
1. d *(112)*
2. b *(113)*
3. c *(113)*
4. a *(114)*
5. c *(114)*
6. a *(114)*
7. c *(114)*
8. d *(114)*
9. c *(115)*
10. a *(115)*
11. c *(115)*
12. b *(115)*
13. b *(116)*
14. d *(116)*
15. a *(117)*
16. d *(117–118)*
17. b *(118)*
18. d *(118)*
19. c *(118–119)*
20. c *(120)*
21. c *(120)*
22. b *(121)*
23. a *(121)*
24. b *(122)*
25. d *(122)*
26. b *(122)*

Identify

C. *(119–120)*
1, 2, 7, 10

Chapter 8 Answers

True/False

A.
1. False. The driver/operator is the *only firefighter* assigned to many ARFF apparatus. *(124)*
2. True *(125)*
3. True *(125–126)*
4. False. The size of dry-chemical systems on ARFF apparatus usually *starts at 500 pounds (227 kg) and can be much larger.* *(126)*
5. True *(126)*
6. False. The future production of *halon extinguishing agents* has been banned because they harm the earth's ozone layer. Some companies have developed *new clean agents* that are not harmful to the environment. *(127)*
7. False. Firefighters *have been killed* from not following directions and improperly servicing clean-agent systems. *(127)*
8. False. ARFF vehicles are typically even larger than structural fire apparatus, and previously trained driver/operators may require additional training. *(127)*
9. False. ARFF vehicles' high center of gravity makes them *prone to rollover.* *(128)*
10. False. The *cost of maintaining vehicle engine, transmission, and braking systems can be high* if driver/operators do not properly operate the ARFF vehicles. *(129)*
11. True *(129)*
12. False. If forced to operate on a steep grade, the ARFF apparatus driver/operator should be *extremely cautious* because the vehicle's *center of gravity definitely changes.* *(130–131)*
13. False. ARFF vehicles are designed to allow for maximum ground clearance, but *this clearance does not prevent* them from catching on something underneath. *(131)*
14. False. ARFF apparatus driver/operators should use *all their senses* when responding in poor visibility, including *slowing down and opening windows to listen* for any strange noises. *(132)*
15. True *(132)*
16. False. The days of hurrying to the aircraft, dumping everything, and driving like crazy back to resupply *are long gone.* *(133)*
17. True *(134)*
18. True *(136)*

Multiple Choice

B.
1. c *(123)*
2. b *(124)*
3. c *(124)*
4. c *(124)*
5. a *(125)*
6. d *(125)*
7. b *(125)*
8. a *(125)*
9. d *(126)*
10. b *(126)*
11. c *(126)*
12. b *(127)*
13. c *(128)*
14. a *(128)*
15. d *(128)*
16. a *(128–129)*
17. b *(129)*
18. d *(129)*
19. b *(130)*
20. c *(130)*
21. b *(131)*
22. d *(131)*
23. a *(131)*
24. a *(132)*
25. b *(133)*
26. c *(133)*
27. d *(133)*
28. b *(134)*
29. d *(134)*
30. a *(135)*
31. d *(135, 136)*
32. c *(136)*
33. b *(136)*
34. c *(136, 137)*
35. a *(137)*

Chapter 9 Answers

Matching

A.
1. i *(144)*
2. m *(145)*
3. b *(145)*
4. h *(145)*
5. j *(145)*
6. c *(145)*
7. l *(145)*
8. g *(145)*
9. n *(146)*
10. d *(146)*
11. k *(147)*
12. a *(150)*
13. f *(156)*

True/False

B.
1. True *(140)*
2. True *(141)*
3. False. Blended fuels such as Jet-B have a *lower* ignition temperature than Jet-A fuels, which makes them potentially more dangerous when spilled. *(142)*
4. False. When using water to extinguish aircraft fires, firefighters have been most successful when they have used *fog or spray streams*. *(143)*
5. True *(144)*
6. False. Although technological advances have somewhat simplified the use of foam, it *is not foolproof*. *(144)*
7. False. Class B foams designed solely for hydrocarbon fires will *not* extinguish polar solvent fires *regardless of the concentration* at which they are used. *(145)*
8. False. Using a foam proportioner that is not compatible with the delivery device (even if the two are made by the same manufacturer) *can result in an unsatisfactory foam* or no foam at all. *(146)*
9. True *(148)*
10. True *(148)*
11. True *(149)*
12. False. Unignited spills *do not require the same application rates* as ignited spills because radiant heat, open flame, and thermal drafts do not attack the finished foam as they would under fire conditions. *(150)*
13. True *(151)*
14. False. Regular protein foam is *no longer* widely used in aircraft fire fighting because it *is corrosive* and *is not self-sealing*. *(151)*
15. False. The in-line eductor is the most common type of foam proportioner used in *structural* fire fighting; however, it *is not commonly used* in ARFF applications. *(152)*
16. True *(152)*
17. False. Apparatus-mounted foam proportioning systems are *commonly* found on structural, industrial, and wildland apparatus and on fire boats *as well as on ARFF apparatus*. *(153)*
18. True *(154)*
19. True *(156)*
20. False. Structural apparatus assigned to an airport facility may have high-energy foam systems, but these systems *are not commonly found* on other types of ARFF apparatus. *(156)*
21. True *(157)*
22. True *(157)*
23. False. The initial foam application at an aircraft fire should be to *insulate the fuselage and protect the integrity of the aircraft skin*. *(159)*
24. False. The terms *dry chemical* and *dry powder* are *often incorrectly used* interchangeably. Dry-chemical agents are for use on Class A-B-C fires and/or Class B-C fires; dry-powder agents are for Class D fires only. *(160–161)*
25. True *(161)*
26. True *(161)*
27. False. Multipurpose A:B:C-rated dry-chemical extinguishers can be used for extinguishing aircraft engine fires but *are not recommended* due to the corrosiveness of the agent. *(161)*
28. False. Wheeled dry-chemical extinguishers store the extinguishing agent and the pressurizing gas in *separate tanks*. The pressurizing gas must be introduced into the agent tank and allowed a few seconds to fully pressurize the tank before the nozzle is opened. *(162)*
29. True *(162)*
30. True *(163)*

9

Multiple Choice

C.

1. b *(140)*
2. c *(140)*
3. a *(140–142)*
4. b *(141)*
5. d *(142)*
6. b *(142)*
7. a *(142)*
8. d *(142)*
9. a *(142)*
10. a *(143)*
11. d *(143)*
12. a *(143)*
13. b *(143)*
14. b *(144)*
15. b *(145)*
16. c *(145)*
17. a *(146)*
18. a *(146)*
19. c *(147)*
20. b *(148)*
21. d *(149)*
22. a *(149)*
23. c *(150)*
24. c *(150)*
25. b *(150)*
26. d *(150)*
27. d *(151)*
28. c *(151)*
29. a *(151)*
30. c *(152)*
31. b *(152)*
32. d *(152)*
33. b *(153)*
34. d *(153)*
35. c *(153)*
36. d *(154)*
37. b *(154)*
38. a *(154)*
39. c *(154)*
40. a *(154–155)*
41. d *(155)*
42. d *(156)*
43. a *(156)*
44. b *(157)*
45. b *(157)*
46. d *(157)*
47. c *(157–158)*
48. c *(158–159)*
49. b *(159)*
50. b *(159)*
51. d *(159)*
52. a *(159)*
53. a *(159)*
54. d *(160)*
55. b *(160)*
56. a *(160)*
57. b *(161)*
58. d *(161)*
59. c *(161)*
60. d *(162)*
61. d *(163)*
62. d *(163)*
63. b *(163)*

Identify

D.

1. Aqueous film forming foam *(150)*
2. Protein foam *(151)*
3. Fluoroprotein foam *(151)*
4. Film forming fluoroprotein foam *(151)*
5. Compressed-air foam system *(156)*

Chapter 10 Answers

True/False

A.
1. True *(167)*
2. True *(167)*
3. False. Apparatus *should not be driven* into gullies and downslope depressions near the aircraft into which fuel may have drained or in which fuel vapors may have collected. *(168–169)*
4. False. If evacuation has begun from the interior of the aircraft, ARFF personnel should *protect the exits being used* and may be tasked to *assist occupants* from the escape slides and direct them to safety. *(169)*
5. True *(171)*
6. False. When a good tire is heated, the increase in air pressure alone is *not* enough to cause it to fail. More likely, the *combination of increased temperature of the brake/wheel assembly and the increased tire pressure* will lead to the disintegration of the wheel assembly. *(172)*
7. True *(172)*
8. True *(174)*
9. False. *Small spills involving an area less than 18 inches (450 mm)* contain such a small amount of fuel that they may be absorbed, picked up, and placed easily in an approved container. *(175)*
10. True *(175)*
11. False. The severity of the hazard created by a fuel spill depends primarily upon *how volatile the fuel is and its proximity to sources of ignition*. *(175–176)*
12. False. It is extremely important that any cargo, baggage, mailbags, or similar items that have come in contact with fuel should be *decontaminated before being placed on board any aircraft*. *(176)*
13. True *(177)*
14. True *(177)*
15. False. Because a free-burning fire in an aircraft almost always vents itself by burning through the skin of the aircraft in the early stages of the fire, *backdraft is unlikely*. *(178)*
16. False. Allowing occupants to exit the aircraft *does not prevent* firefighters from opening all available exit doors, hatches, and windows in an attempt to ventilate the aircraft. *(178)*
17. False. *Other than radioactive materials*, dangerous goods must be accessible by the flight crew and are often stored near the front of the aircraft. *(179)*
18. False. When a fire warning light activates while in flight, the crew *tries to determine whether there is a fire* by making instrument checks and visual observations. *(180)*
19. True *(181)*
20. True *(182–183)*
21. True *(183)*
22. False. Promptness and *safety are equally important* response considerations. *(184)*
23. True *(185)*
24. False. Emergencies that occur without prior warning are called *unannounced emergencies*. *(185)*
25. False. At times, *it is difficult to distinguish* between rescue and extinguishment activities because they are interrelated and are often performed simultaneously. *(185)*
26. True *(186)*
27. False. In aircraft accidents involving fire, the initial attack generally involves operating both roof and bumper turrets *while additional units perform handline operations*. *(186)*
28. False. Cutting ventilation openings in an aircraft fuselage *is not recommended* because it is time-consuming, may be dangerous, and should be considered as a last resort. *(186)*
29. False. Rescue apparatus should be positioned so that all rescue and forcible entry equipment is as near to the *probable point of entry* as possible without endangering the vehicle. *(188)*
30. True *(189)*
31. False. Whenever possible, using an *extrication board or backboard is the preferred method* of supporting a person with a suspected back injury. *(189)*
32. False. A thorough overhaul inspection at aircraft incidents is necessary *whether fire was apparent or not*. *(190)*
33. True *(191)*
34. False. Many military aircraft use a varied mixture of jet fuel, which has a significantly *lower* flash point than civil aviation fuel. *(191)*
35. False. An area should be established where the walking wounded can be taken in order to prevent them from returning to the crash site, to isolate them from the scene for mental health reasons, and *to ensure that they are not bothered by members of the press or attorneys*. *(193)*

Multiple Choice

B.

1.	b *(167)*	32.	b *(179)*
2.	b *(167)*	33.	a *(179)*
3.	d *(167)*	34.	d *(180)*
4.	c *(168)*	35.	d *(181)*
5.	a *(169)*	36.	c *(181)*
6.	d *(169–170)*	37.	d *(181)*
7.	c *(170)*	38.	d *(182)*
8.	a *(170–171)*	39.	a *(182)*
9.	b *(171)*	40.	b *(182)*
10.	a *(171)*	41.	a *(183)*
11.	d *(171)*	42.	c *(183)*
12.	b *(172)*	43.	a *(183)*
13.	c *(172)*	44.	b *(184)*
14.	c *(172)*	45.	c *(185)*
15.	a *(173)*	46.	c *(185)*
16.	d *(173)*	47.	a *(185)*
17.	c *(173)*	48.	d *(186)*
18.	b *(173)*	49.	d *(186)*
19.	a *(173)*	50.	b *(188)*
20.	c *(174)*	51.	c *(189)*
21.	b *(174)*	52.	d *(189)*
22.	c *(174)*	53.	a *(189)*
23.	b *(175)*	54.	d *(190)*
24.	a *(175)*	55.	d *(190)*
25.	d *(175)*	56.	a *(190)*
26.	a *(176)*	57.	b *(191)*
27.	c *(176)*	58.	a *(191)*
28.	d *(177)*	59.	c *(191)*
29.	b *(178)*	60.	c *(192)*
30.	b *(178–179)*	61.	b *(192)*
31.	b *(179)*	62.	c *(193*

Identify

C. *(167–168)*
4, 5

D. *(174)*
1, 3, 4, 7

E. *(175)*
1, 2, 4

Chapter 11 Answers

True/False

A.
1. False. Developing an airport emergency plan is *not* an end in itself, nor is it a guarantee for an effective emergency response *(195)*.
2. True *(196)*
3. True *(196)*
4. False. ARFF personnel handle accidents/incidents involving general aviation aircraft *in the same way* they do those involving commercial aircraft. *(197)*
5. False. All concerned with aircraft emergency services must have grid maps of the airport and surrounding areas within a *5- to 15-mile (8 km to 24 km)* radius. *(199)*
6. True *(199)*
7. True *(201)*
8. False. Nonemergency vehicles *may be used* to transport *ambulatory or uninjured* persons. *(201)*
9. False. The emphasis in patient care following most aircraft accidents is on *immediate medical stabilization* of the injured and *timely transport* to the nearest medical facility. *(201)*
10. True *(202)*
11. True *(203)*
12. False. To reduce anxiety among relatives and friends of occupants of an involved aircraft, the investigative authorities may authorize release of the names of those not seriously injured *as soon as possible*. *(204)*
13. True *(204)*
14. True *(205)*
15. False. Military police have *no peace-officer authority* unless a military aircraft incident is declared a National Defense Area. *(206)*

Multiple Choice

B.
1. c *(196)*
2. a *(196)*
3. d *(196)*
4. b *(197–198)*
5. a *(198)*
6. b *(198)*
7. c *(199)*
8. d *(199–200)*
9. c *(200)*
10. a *(200)*
11. c *(200)*
12. b *(201)*
13. c *(201)*
14. d *(201–202)*
15. c *(202)*
16. b *(202–203)*
17. a *(203)*
18. b *(203)*
19. a *(203–204)*
20. d *(204)*
21. c *(204–205)*
22. b *(205)*
23. a *(206)*
24. d *(206)*
25. c *(207)*
26. c *(207)*
27. c *(208)*
28. d *(208)*

Identify

C.
1. Airport emergency plan *(195)*
2. National Transportation Safety Board (U.S.) *(204)*
3. Canadian Transportation Accident Investigation and Safety Board *(204)*
4. Public information officer *(205)*
5. Federal Emergency Management Agency (U.S.) *(196, 206)*
6. National Defense Area *(206)*
7. Explosive ordnance disposal *(206)*

D. *(205–206)*
2, 3, 5, 7

E. *(207–208)*
1, 4, 5, 6, 8

Chapter 12 Answers

Matching

A.
1. c *(211)*
2. e *(211)*
3. h *(211)*
4. g *(211)*
5. a *(211)*
6. d *(211)*
7. f *(211)*
8. i *(211)*
9. b *(211)*

B.
1. c *(215)*
2. d *(215)*
3. b *(215)*
4. a *(215)*

True/False

C.
1. False. *Hazardous materials* is a United States *fire service term*. In the aviation industry, the term *dangerous goods* is used. *(209)*
2. True *(210)*
3. False. These requirements *do not guarantee* that illegal "undeclared" shipments will not be on board an aircraft. *(210)*
4. False. Certain hazardous cargoes such as dry ice and magnetized materials *may not always* be stowed in specialized containers but may be stowed in *any aircraft cargo compartment.* *(213)*
5. False. Because of the wide variety of circumstances in which dangerous goods may be encountered, their identifications *may be challenging* in air transport situations. *(213)*
6. True *(213)*
7. False. Once the materials involved are *identified and verified and all their properties and characteristics are known,* a mitigation plan may be devised for the disposition of the problem. *(214)*
8. True *(215)*
9. True *(215)*
10. False. Because ARFF personnel will probably be involved in dangerous goods mitigation or other essential activities in and around the aircraft, large-scale evacuations probably will become the responsibility of *law enforcement or other personnel.* *(216)*
11. False. While ARFF personnel may be placed at some degree of risk to effect rescues, their being put in jeopardy to recover bodies *is inappropriate.* *(216)*

Multiple Choice

D.
1. c *(209–210)*
2. a *(210)*
3. d *(210)*
4. c *(211)*
5. b *(211)*
6. c *(211)*
7. a *(212)*
8. c *(212)*
9. a *(213)*
10. c *(213)*
11. d *(213)*
12. b *(214)*
13. b *(214)*
14. a *(214)*
15. c *(214)*
16. d *(215)*
17. d *(215)*
18. b *(215)*
19. b *(216)*
20. d *(216)*
21. c *(216)*
22. a *(216)*

Identify

E.
1. Dangerous goods *(209)*
2. International Air Transport Association *(210)*
3. Reportable quantities *(210)*
4. *Chemical Hazard Response Information System (214)*
5. Bureau of Explosives *(214)*
6. Association of American Railroads *(214)*
7. Chemical Transportation Emergency Center *(214)*
8. Canadian Transport Emergency Centre *(214)*
9. Hazardous materials response team *(215)*
10. Environmental Protection Agency (U.S.) *(215)*
11. Structural firefighter protective clothing *(215)*

Notes